Stellar Alchemy
The Celestial Origin of Atoms

Why do the stars shine? What messages can we read in the light they send to us from the depths of the night? Nuclear astrophysics is a fascinating discipline, and enables connections to be made between atoms, stars and human beings. Through modern astronomy, scientists have managed to unravel the full history of the chemical elements, and to understand how they originated and evolved into all the elements that compose our surroundings today. The transformation of metals into gold, something once dreamed of by alchemists, is a process commonly occurring in the cores of massive stars. But the most exciting revelation is the intimate connection that humanity itself has with the debris of exploded stars. This engaging account of nucleosynthesis in stars, and the associated chemical evolution of the Universe, is suitable for the general reader.

MICHEL CASSÉ is an astrophysicist at the Service d'Astrophysique in Saclay, France, and an associate research scientist at the Institut d'Astrophysique de Paris. He has published several other popular physics books in French.

This book was translated by Stephen Lyle, a freelance translator of physics and astrophysics.

Stellar Alchemy
The Celestial Origin of Atoms

MICHEL CASSÉ
Service d'Astrophysique
and Institut d'Astrophysique de Paris

Translated by Stephen Lyle

CAMBRIDGE
UNIVERSITY PRESS

PUBLISHED BY THE PRESS SYNDICATE OF THE UNIVERSITY OF CAMBRIDGE
The Pitt Building, Trumpington Street, Cambridge, United Kingdom

CAMBRIDGE UNIVERSITY PRESS
The Edinburgh Building, Cambridge CB2 2RU, UK
40 West 20th Street, New York, NY 10011-4211, USA
477 Williamstown Road, Port Melbourne, VIC 3207, Australia
Ruiz de Alarcón 13, 28014 Madrid, Spain
Dock House, The Waterfront, Cape Town 8001, South Africa

http://www.cambridge.org

First published 2003

Printed in the United Kingdom at the University Press, Cambridge

Typeface Times 10/13 pt   *System* LaTeX 2_ε [TB]

A catalogue record for this book is available from the British Library

Library of Congress Cataloguing in Publication data

Cassé, Michel, 1943–
Stellar alchemy : the celestial origin of matter / Michel Cassé.
p.   cm.
Includes bibliographical references and index.
ISBN 0-521-82182-7
1. Nuclear astrophysics.   I. Title
QB463.C37   2003
523.01'97 – dc21   2002041451

ISBN 0 521 82182 7 hardback

This book is dedicated to André Grau

To exorcise the sadness of the finite,
I have chosen the sky.

Always polite with light,
I let it have the first word.

Instruction should be gentle.

Contents

Preface

Rising ideas in the sky of knowledge

The constellated sky has never ceased to foster the enthusiasm of men and women in search of illumination. In his *Theogony*, Hesiod tells us how Gaea, the wide-bosomed Earth, arose from the vast and dark Chaos, accompanied by Eros. Gaea first bore Uranus, the starry heaven, then the barren waters of the sea.

Almost three millennia separate this genealogy of the gods from the modern idea of the expanding cosmos and the stellar origins of human matter, summed up so concisely in the two statements:

* the Universe is expanding;
* we are made from star dust.

What brought human beings to invent cosmology and relegate the celestial genealogy of the gods to oblivion? Is there no more drama in the heavens? No genesis?

The stars are just the punctuation in the text of the heavenly narrative. And yet, in our recitation, we claim to know the whole history of the Universe. Formulating its plot on the stage provided by the space–time of matter, astrophysics expresses a cosmic vision as well as a body of scientific thought. The sky is a palimpsest. Under the first visible writing, the starry sky, modern astronomy has managed to bring out, at least in part, very ancient hieroglyphics and original engravings.

From Hesiod to Aristotle, from Aristotle to Galileo, from Galileo to Einstein, the Universe has undergone repeated mental reform. Visionaries of infinity have replaced poets and men of God. Byzantine mosaics and West European stained glass portrayed theories of the world. Today, equations between symbols have replaced the liturgies, but the quest for unity stands.

This book contains a defence and illustration of nuclear astrophysics, one of the most beautiful sciences there is. It bridges the gap between the atomic microcosm and the celestial macrocosm, setting out the origin and evolution of all the elements that make up our immediate surroundings. In the most harmonious way, it combines the physics of the very small and the very large, the inner workings of nuclei at one end of the scale and stars at the other. This sumptuous marriage between nuclear physics and astronomy, Earth and sky, celebrated in scientific thought, opens the way to a genuinely universal history of the material substance constituting all visible things.

It is indeed an admirable science that studies the genesis and evolution of the chemical elements composing nature and establishes a historical relationship between all the forms in the sky: the primordial light, particles, atoms, stars and human beings. Much lyricism has been inspired. Let us cite: "Without knowing it, we have been working with star dust, carried to us by the wind, and we drink the Universe in a raindrop." Or again: "We are ash and dust, that may be true, but we are the ash and dust of stars." This qualitative and poetic, but emotional and even violent, face of astrophysics is the one I lean towards. But fitted out with dove's wings, it is threatened with hysteria, in the sense that Baudelaire is a hysterical Boileau. This is not the one I wish to speak about here.

The aim of the book is to describe in all simplicity the combinatorial and quantitative science of nucleosynthesis, a pure stellar arithmetic, chaste and contemplative, cloaked in the ironic reticence of numbers. It is so penetrating and revealing that it may be considered as our most powerful tool for divulging the material history of the Cosmos.

The sciences of the Universe contribute to humanism, education and the advent of the technological society. The greatest contribution of fundamental astrophysics is to answer questions about the place of humanity in the cosmos. Quantitative answers can be given to age-old questions, some of which border on the metaphysical or even the religious.

Not content to satisfy immemorial curiosity about the cosmos, the sciences of space and the Universe have nurtured a hotbed of technological innovation which has not failed to influence our daily lives. Technical spin-offs from space research have been a boon to industry, medecine and the environmental sciences.

Industry invests in space science and has earned significant dividends. The return on investment is considerable, primarily in its financial and technical guise, but also, in a less direct way, through formal education as administered in schools, colleges and universities, and informal education in the form of television, newpapers and magazines, exhibitions, lectures and planetariums. The study of the heavens encourages a younger public to develop quantitative reasoning and, in the case of amateur astronomers, for example, to contribute to the setting-up of concrete and active observation strategies.

By their very definition, the sky sciences are an open area. It was international partnerships within the European astrophysical community that opened the way to economic and political unification. Indeed, Europe began in the sky, with the foundation of the European Southern Observatory (ESO) and the European Space Agency (ESA), and today, astrophysical conferences have become planetary events. We can only rejoice!

Acknowledgements

I would like to express my warmest thanks to Odile Jacob for her wisdom and solar obstinacy, and to Gérard Jorland, Jean-Luc Fidel and Aurèle Cariès who brought the shine back to the star of literature. I would also like to pay homage and compliments to Elisabeth Vangioni-Flam, a bright light who scolds the Big Bang. My thanks go to Jean Audouze, Jean-Claude Carrière, Matta, Jacques Paul, Catherine Cesarsky, Alfred Vidal-Madjar and Roland Lehoucq for their stellar fraternity.

The publisher has used its best endeavours to ensure that the URLs for external websites referred to in this book are correct and active at the time of going to press. However, the publisher has no responsibility for the websites and can make no guarantee that a site will remain live or that the content is or will remain appropriate.

1

Nuclear astrophysics: defence and illustration

Glossary

baryons heavy particles subject to the strong interaction
Big Bang initiating event in cosmology
cosmic rays high speed particles moving across the Galaxy
cosmological background radiation electromagnetic relic from the primordial Universe
cosmological constant term added by Einstein to his cosmological equations to obtain a static model of the Universe
degenerate term used in quantum physics to designate a fluid whose density is such that the electrons within it refuse to occupy a lesser volume and consequently exert a constant pressure which opposes further contraction
gamma rays high-energy electromagnetic radiation
hydrogen fusion chain of nuclear reactions leading to the production of helium
kilo electron volt (keV) 1000 electron volts, where 1 eV is the energy acquired by an electron in a potential difference of 1 volt
neutrino extremely light particle insensitive to the strong interaction
neutron electrically neutral particle making up atomic nuclei
neutron star very compact star composed mainly of neutrons
nucleon proton or neutron
nucleosynthesis production of atomic nuclei in the Universe
quintessence substrate exerting a gravitational repulsion
scalar field a non-directional type of field required in cosmology and particle physics
supernova stellar explosion
Wolf–Rayet star massive star producing a high stellar wind

Aims and aspirations of nuclear astrophysics

The aim of nuclear astrophysics is threefold. Firstly, it seeks to determine the *mechanisms* whereby the various nuclear species occurring in nature are built up, from deuterium with its two nucleons to uranium with 238 nucleons. Secondly, it seeks to identify the astrophysical *site* in which these species are produced. And thirdly, it attempts to unravel the *temporal sequence* of the nuclear phenomena that fashion baryonic matter, the stuff of stars and humans, making up the galaxies. Beyond this, it aims to explain the composition of the Solar System and the main trends of chemical evolution in the Galaxy, such as the gradual enrichment in metals and the relative abundances of the elements. It explores in detail everything from the first stages of chemical evolution in the Universe through to the most recent and violent events of nucleosynthesis related for the main part to supernovas and large stars with strong stellar winds (Wolf–Rayet stars). Indeed, gamma photons of precise energies are emitted by the freshly fashioned nuclei in high-wind stars and supernovas, radioactive nuclei in search of their ultimate form. Without a doubt, these high-energy photons constitute the purest clues as to the mechanisms producing atomic nuclei in the Universe. Hence, the spectroscopy and mapping of celestial gamma sources should provide us with fresh evidence of nucleosynthesis and locate its centres in our own galaxy and beyond.[1]

One of the great achievements that future generations will associate with the twentieth century is the understanding of the mechanism that makes the Sun shine. Indeed, it is the same mechanism that operates in all stars. It is thus established that the stars are not eternal. With this comes the realisation that we ourselves are but the dust and ashes of stars and that we have borrowed the elements that make us up from these celestial workings.

The origins of the atomic nuclei from hydrogen to uranium have been carefully established: the Big Bang in the case of hydrogen, helium and a dash of

---

[1] Strange, dedicated space-borne telescopes have been launched to catch nucleosynthesis in the act, by mapping celestial gamma emissions. This is not the place to go into the particular relationships that have formed between space technology, fundamental astrophysics and the delicate matter of detecting photons with energies greater than 100 keV which are the bread and butter of gamma-ray astronomy. A useful reference is *The Universe in Gamma Rays* edited by Volker Schönfelder (Springer-Verlag, Berlin, 1998). The theoretical aspects of nuclear gamma astronomy are laid out in my own lecture notes *Nucléosynthèse et abondance dans l'univers* (Cépaduès Editions, Paris, 1998), written in the context of the twentieth spring school of astrophysics at Goutelas.

After an initial phase dominated by the United States, nuclear gamma astronomy should soon enter the fold of the Old Continent. The key factor in European strategy is the satellite INTEGRAL (International Gamma Ray Laboratory), developed under the auspices of the European Space Agency at the Service d'astrophysique in Saclay, France and the Centre d'Etude Spatiale du Rayonnement in Toulouse, France, among others. This telescope was launched in 2002. Its spectroscopic observations should bring answers to some of the most fundamental questions relating to stellar and interstellar nucleosynthesis.

lithium-7, the stars for carbon to uranium, and cosmic rays for lithium, beryllium and boron.

By the 1990s, nuclear astrophysics was no longer suspended in theoretical limbo. It had ceased to be a purely speculative discipline and was becoming a science in its own right, with its own quantitative analysis and predictions, open to confirmation by physical measurement. For example, the measured level of the solar neutrino flux, even though it does not coincide in detail with the expected value, attests to the truth of the basic ideas of nuclear astrophysics as applied to our own Sun, and by extrapolation, to all stars. (The discrepancy is roughly a factor of two, and this can very likely be put down to problems in our understanding of neutrinos, rather than some defect in stellar physics.) More precisely, it corroborates our ideas about hydrogen fusion in low-mass stars, a key phase in their lives, allowing them to shine for billions of years.

At the other extreme, the neutrinos and gamma rays emitted by a supernova in 1987 brought brilliant confirmation of the most elaborate theoretical speculations concerning the mechanisms of stellar explosion and the gravitational collapse which immediately precedes it.

Over a period of forty years, the theory of stellar nucleosynthesis has been transformed from an abstract description of the various nuclear processes that fashion matter to a fully evolved discipline, quite able to stand up to confrontation with a vast body of observational data.

The future is likely to be no less rewarding. Astronomers have not let success go to their heads and their thirst for discovery and knowledge has remained intact. They are certainly not so naive as to believe that the stars have delivered up all their secrets. The most ancient stars, those born before the Galaxy had assumed its present form, have now become a subject of intense interest. The next goal on the distant horizon is a complete picture of chemical evolution in space. In this context, it is quite clear that the early stages of this evolution are the least well understood. The end of the road is not yet in sight.

An historical overview

We must observe the sky with the right theory. 'I see' means 'I understand'. 'Clarity comes from thought.' By close observation, I can alter my intellect. One night in 1572, Tycho Brahe spotted a new star in the heavens and abolished at a glance the dogma asserting the eternity and perfection of the superlunary sky, smashing Aristotle's crystalline spheres to smithereens. He concluded quite rightly that the upper regions of the sky, beyond the Moon, themselves belong to the sphere of birth and corruption. They are neither inalterable nor impassive. However, since he referred to this as a new star, or nova, and indeed he was so struck by it that he wrote a book *De stella nova*, he did in fact make a mistake. For this was not one more star in the heavens, but one less star! This type of phenomenon, known today as a supernova, is the luminous expression of an explosion of a massive star or a binary star system.

No indeed! The stars are not made from some incorruptible ether but of ordinary matter. The last post had sounded for the quintessence, the fifth and highest of the elements, along with earth, air, fire and water. Tycho enlisted the services of Kepler and in 1604 master and pupil were rewarded by the magnificent spectacle of a supernova.

The telescope had not yet appeared. This invention would have to wait for Galileo to arrive on the scene. In 1610, pointing his newly created instrument towards the Moon, Galileo saw mountains. From this, he inferred that it was Earth-like. Today, we would be more inclined to say that the Earth is celestial, as I myself like to reiterate from one book to the next, for I find the fact so startling. At this moment, the distinction between the sublunary sphere, seat of transformation and death, and the superlunary sphere, eternal and perfect, was forever erased. And once the Earth was recognised to be celestial, the sky became mortal and comprehensible.

The discovery that stars were transitory and the Moon a material object served to abolish the crystal barrier between heaven and Earth in the minds of Renaissance astronomers. Astrophysics was born, proclaiming the material nature of the celestial bodies, and the whole sky suddenly became intelligible. Earth-bound observers began to convince themselves that time brings the end of all things, including the stars, and hence that the stars were themselves things. The quintessence or ether was gradually to fade away, leaving the way open for atoms.

The first equation of astrophysics is thus: Earth = sky. What is atom, light and physical law down here is just the same up there. What is like nothing down here is like nothing anywhere. The world is made up of atoms, disposed so as to form the Universe.

It was Newton, riding on the shoulders of giants as he liked to say when referring to his predecessors, who provided a permanent foundation for the physics of the heavenly bodies. The same gravity holds the Moon and causes the apple to fall. The new heavens now had their constitution.

This marriage between Earth and heaven, made in human thought, was now consummated. From this moment, the science of the heavens would never cease to prosper. Physics had given astronomy a head and astronomy would give physics wings.

Humanity has never been indifferent to heavenly signs. Any change in the appearance of the night sky, visible to the naked eye, has always led human beings to speculation, lamentation, prediction and, in recent times, explanation.

As far as we know, the oldest recording of a stellar explosion (at least, this is the way it is usually interpreted) was engraved on an animal bone by someone in China in 1300 BC: 'On the seventh day of the month, a great new star appeared in the company of Antares.' The Sung Dynasty may proudly claim to have recorded three supernovas in its celestial registers, dating to AD 1006, 1054 (a famous event) and 1181.

Although 'new stars' (novas and supernovas) had been observed for many centuries, as can be seen from Chinese, Japanese and Korean chronicles, the modern, scientific age of supernovas only began on 31 August 1885, when Hartwig discovered a new star near the centre of the Andromeda galaxy. Eighteen months later the new star had disappeared.

In 1919 Lundmark estimated the distance to the Andromeda galaxy at some 700 000 light-years. (Current estimates put it at 2 million light-years.) It was clear that Hartwig's star was a thousand times brighter than any known nova. It was the same Lundmark who suggested a connection between the supernova observed by Chinese astronomers in 1054 (the year of the religious schism) and the Crab nebula.

An event similar to the Andromeda supernova was observed in 1895 in the galaxy NGC 5253. This time the new star grew brighter than its host galaxy.

It was not until 1934 that Baade and Zwicky succeeded in making a clear distinction between the classic and commonly observed novas (occurring at a rate of about 400 per year in our own galaxy) and the much rarer supernovas. Fred Zwicky's systematic searches between 1956 and 1963 resulted in the discovery of 136 supernovas.

When it became possible to obtain the spectrum of one of these objects in 1937, it was obvious that they looked like nothing yet known. All supernovas discovered in subsequent years displayed a remarkable uniformity, both in intensity and in behaviour. This observation led Zwicky to suggest that they might be used as standard candles to calibrate distance across the cosmos. But then, in 1940, a supernova with a completely different spectrum was discovered. It soon became clear that there were at least two classes of supernova, distinguished by their spectral features. It was the presence or absence of the Balmer lines of hydrogen near the maximum of the light curve that provided this classification.

A scenario imagined by Zwicky in 1938 was for a long time the only explanation of the phenomenon. According to this view, a supernova marks the transformation of a normal star into a neutron star, drawing its energy from gravitational collapse. This led astronomers to think that the death of a star was the transition from luminous perfection to a kind of dark perfection.

In 1960 Fred Hoyle and William Fowler discovered that thermonuclear combustion in the dense core of a degenerate star (the word 'degenerate' is used in the sense of quantum theory and will be made explicit later) could trigger the explosion and volatilisation of the star. If we add the idea that *post mortem* light emissions are fuelled by the gradual disintegration of an unstable radioactive isotope, nickel-56, a subject to be discussed in great detail later, we obtain the universal explanation of what are now known as type Ia supernovas.

For over four centuries, we were deprived of any nearby stellar explosion. It was not until 1987 that a supernova visible to the naked eye finally graced the stage. It made its appearance in the Large Magellanic Cloud, a modest galaxy gravitating around our own Milky Way at a distance of some 170 000 light-years. All suitably placed instruments of observation available at the time, both on Earth and in the sky, were pointed towards the exploding star. The gradual decline in its brightness and the slightest fluctuation in its level of radiation were monitored by observatories across the southern hemisphere (for it was not visible from northern latitudes). Gigantic neutrino detectors hidden deep under the Earth's surface were also waiting to pick up its signal. X and gamma rays were recorded by dedicated satellite-borne detectors. It was a unique opportunity to put the theory of stellar explosion and explosive nucleosynthesis to the test – and the theory was confirmed. Humankind had pursued this dream of the stars by calculation, from their birth in gigantic clouds to their death in fleeting light, and with it the genesis of the elements and the celestial extraction of humankind itself.

Apart from the welcome news of a celestial origin for all atomic matter, the supernovas were to reserve an unexpected bonus, revealing perhaps more clearly something that other signs were already trying to tell us: matter is not alone in the Universe, but has a sister, which has just been christened somewhat playfully quintessence. Of illustrious birth, for it was already present in the first moments of the Universe, this invisible sister weighs in at a good fifteen times its atomic counterpart. Paradoxically, however, it exerts nothing other than a repulsive gravitational effect! How was the existence of this new ether brought to light? By observing distant supernovas and applying the arguments of modern cosmology (see Appendix 2).

As is well known, the expansion of the Universe was revealed through the redshifting of light from distant galaxies. The fate of such a dynamic universe then became one of the main questions to be addressed by cosmology.

Today, it is generally thought that the Universe is open, by which it is meant that the expansion is eternal, and indeed, even accelerating. This strikes at the very heart of the old paradigm of eternal return. This announcement of a one-way diaspora originates from a certain type of exploding star, visible from the very depths of the cosmos. Indeed, the type Ia supernovas are so bright and so regular in their characteristics that they can be used for calibration purposes, as so-called standard candles. Recent systematic studies of such objects have revealed, against all expectations, a slight weakening of the light flux which in turn attests to a slightly greater stretching of space than had been previously predicted. From this small increase in the expansion relative to the results of the canonical calculation, it can be deduced that the expansion is actually accelerating. The cause of this acceleration is referred to, tongue-in-cheek, as quintessence. It produces a repulsive gravitational effect in the same way as the cosmological constant introduced by Einstein to prevent the Universe from collapsing. And so the notion of matter has been extended. The irony of the situation is that, although quintessence has made a comeback, it is really just a word to cover up our astronomical ignorance.

Physicists in their all-pervading seriousness prefer the term 'scalar field' to 'quintessence'. For the quintessence field is indeed indifferent to direction, like a mere number. Space has no preferred axis just as it has no preferred centre.

The scalar field that is the object of their greatest desires and most ardent research is called the Higgs field. It arises as a logical necessity in their grand unification of the forces of nature. However, the cosmic scalar field mentioned above is of a very different nature, for it is infinitely lighter than the Higgs field.

When American astronomers Riess, Perlmutter and colleagues proclaimed this acceleration of the expansion of the Universe against a background of distant supernovas, a heated debate began amongst the world's cosmologists. It is now important to check that evolution of supernovas could not give rise to comparable effects, that is, that it could not provide an alternative explanation to the cosmological constant, or the quintessence, its substitute as stretcher of space. It must be shown by both calculation and observation that distant, metal-poor supernovas are not less luminous than closer ones.

The expansion of the Universe is accelerating! The news is too recent to accept without reserve. It must be checked again and again, in the most critical spirit, and yet with open mind, for it is of the utmost importance.

Explosive astrophysics

Every region of the Universe is evolving, but the most spectacular evolution concerns its geometry. Space is expanding between clusters of galaxies. However, this cosmic picture, no matter how generous it may appear, is still far too abstract. The question of the materiality of the Earth and the sky is left unanswered. Where is the world's flesh? The search for the material origins of the elements can bridge the gap between the Universe as it appears on paper, in the stark form of a space–time diagram, and the world incarnate in which the same patterns emerge time and time again: stars in the sky and atoms in the stars. Astrophysics places humankind in space and time.

Supernova! Stellar explosion followed by the decay of radioactive nuclei. A whole generation of *Homo astronomicus* is now involved in the physics of such cataclysms. These spectacular events are now the subject of such intense scrutiny that one might say that humanity has entered the age of the supernova.

They have been thrust to the forefront of cosmology, a boon for nuclear astrophysics! The satellite INTEGRAL, dedicated to nuclear gamma-ray astronomy and on the point of spreading its wings, carries all the hopes and aspirations of the scientific community to which I belong.

Future advances may be expected from the following facilities, either separately or in combination:

- development of optical spectroscopy at the VLT (Very Large Telescope) and elsewhere;
- development of spectroscopy outside visible wavelengths, with ALMA– FIRST (submillimetre and infrared), FUSE (ultraviolet), XMM and Chandra (X ray), INTEGRAL (gamma ray), and a whole range of new astronomical instruments;
- enhancement of computational resources, both in terms of software and hardware, as research institutes are equipped with powerful data-processing and simulation systems;
- coordination of national and international research programmes, such as the systematic search for high-redshift supernovas for cosmological purposes;
- availability of new scientific instruments, such as the megajoule laser, a perfect tool for studying explosions and other violent phenomena, or ion probes, used for accurate isotopic analysis of meteorite samples.

In other words, nuclear astrophysics is about to blossom, a prospect I greet with great joy, and yet not without mixed feelings. For despite the long list of stellar and material emotions it excites, and despite the star it hoists in the sky of knowledge, I remain doubtful whether these celestial scientific reflections will penetrate the conscience of humankind. For whilst technology is at the gallop, the human soul makes its way as placidly as ever. The likes of Jean Audouze, Hubert Reeves, Alfred Vidal-Madjar, T.X. Thuan and many others have carried the banner, keen to share this newfound beauty in the sky. Astronomy is an exciting science, working with powerful research tools, and astronomers are generally courteous, sometimes eloquent, but quite clearly less charismatic than the wise men.

The association of stars into figurative groups or constellations was the first form of hieroglyphic writing, engraving exotic and mythical creatures almost indelibly on the firmament of human consciousness (or unconsciousness). So enduring is this image that I cannot but ask myself why the balance so quickly did away with alchemy, when the telescope has proved unable to unmask astrology. I wonder also why the nuclear alchemy of nucleosynthesis accomplished within the stars leaves the horoscope readers cold, as they search in anguish for some secret co-operation between heaven and Earth. For modern science, the link between humankind and stars is more than just symbolic: it is genetic. Today we may speak of a genuine genealogy of matter in which the stars are clearly designated as our ancestors. This knowledge has not come down from philosophers and men of faith, but from engineers and mathematicians.

2

Light from atoms, light from the sky

Glossary

dark matter unseen and possibly invisible matter
pulsar neutron star with a high magnetic field, emitting narrow beams of radiation

Matter with and without light

To begin with, how do astronomers know what stars are made of? The answer is that they have learnt to decode the language of light, of all the different kinds of light, be they visible or invisible. Atoms in stars speak to atoms in eyes using the language of light. I say it now, and I will say it again until the message is clear. It is the identical nature of emitter and receiver that makes perception possible.

Light has become as a mother tongue for astronomers, a conscientious and light-footed messenger, speaking volubly in every region of space, transporting information from one point of the Universe to another. It is like an expressive covering for the atom. Carnal and material beings long believed that all matter was like itself composed of atoms. Materialism merged with atomism and was taught as a definitive doctrine. It is quite understandable that astronomy, science of light, would be tempted to confine the Universe to its visible aspect, imprisoning it somehow in its mere appearance.

Then suddenly, against her will but in all lucidity, Urania the ancient Greek muse of astronomy was compelled to admit that what is visible is only the froth of existence. She even came to give precedence to what cannot be seen, to what neither shines nor absorbs light. Astronomical observation and theoretical reasoning suggest that most of matter is actually invisible.

A great wound was thus inflicted upon the world of thought and light, whilst dark matter and quintessence gradually took possession of physical space. The twentieth century, when the word 'atom' was on everyone's lips, closed with a cosmic declaration of the atom's insignificance. Two all-pervading ectoplasms have changed the face of the sky and the definition of matter. But how did they come to take control? A glance at the vast collection of data gathered by astrophysicists and summarised at the end of this book will soon convince the reader that the new paradigm is based on experiment and reasoning. After the greatest investigation ever carried out into the nature of the skies, we arrive at the following conclusion: the Universe is not dominated by luminous matter in stars, but by a dark ectoplasm.

It is ironic to note that the most fundamental conclusion reached by astronomy after thousands of years of dedicated observation does not concern what can be seen of the cosmos, but rather what cannot be seen: dark matter. A strange achievement indeed! Dark, transparent or invisible, all the forms of luminous neutrality have burst forth upon astronomy, the science of light. Dark matter has no other expression than gravity to relate it to the world, and gravity is nothing but the attraction of matter for other matter. It is in this sense that it constitutes matter in its own right. We can detect its presence by analysing the motions of bright objects held within its grasp, like stars within galaxies and galaxies within clusters of galaxies. We have no choice but to postulate the existence of this matter if we are to explain the motion and confinement of luminous matter. And yet we know nothing of its nature and composition.

The Herculean task of astronomy and cosmology is thus to draw up a complete inventory of the Universe, determine its evolution, and seek out the cause and the driving force. One of the key aims of cosmology is to establish the true relationship between the geometric structure of the Universe and the distribution of matter and energy contained within it. Once we understand the interplay between space, time and energy, it becomes possible to determine a genuine history of the Universe, a history which can be extrapolated into the past and the future to rewrite the genesis and the apocalypse, as it were.

So here comes humankind shouting to the winds that atomic matter, cradle of all that is visible and tangible, comprises only one to four percent of the substance of the Universe. And all that concerns the notion of matter concerns reality itself.

Atomic matter is the blossoming of all matter, making up for its weakness in numbers by its force of expression. For humankind, it is the sensorial manifestation of the Universe, its crowning eloquence. Indeed, it stands out by its expression in light, in stark contrast to dark matter which is totally indifferent to electromagnetic radiation, neither absorbing nor emitting it, in this respect a featureless entity.

Stellar matter exhausts itself to maintain its sparkle, diamond dust in a cold crystal. The galaxies are glitter, sprinkled across the heavy, transparent darkness of the cosmos.

But if the atom is not the only face of matter, as we seem forced to admit, it is to our knowledge the atom that assumes the widest and most evolutive range of forms. Dark matter and its companion ectoplasm, the repulsive substrate of the Universe, transparent yet crammed with energy, that we venture to call 'vacuum' (Cassé 1993) or 'quintessence', are merely shapeless substrates before the atom in its wealth of forms, fine tracery and beauty. And raised above all else, it is cathedrals of atoms that form the seat of life and consciousness, giving speech to matter.

Luminous matter has revealed dark matter, but the new substance remains obscure. What is it made from? Is it perhaps composed of known forms of matter? Only partly! Is dark matter made up of microscopic particles? If the answer is affirmative, we may suppose that this unknown form of energy penetrates and permeates the galaxies, the Solar System and even our own bodies, just as neutrinos pass through us every second without affecting us in any way. And like the neutrinos, these unknown particles would hardly interact at all with ordinary matter made from atoms. To absorb its own neutrinos, a star with the same density as the Sun would have to measure a billion solar radii in diameter. Luminous and radiating matter is a mere glimmer to dark matter.

We must resign ourselves. The simple Greek idea that matter can be reduced to a handful of atoms must be amended. The greater part of the universal substrate is invisible, because it does not radiate. The invisible state of matter exceeds, both in volume and mass, the manifest state, which is luminous and legible. Atomic matter made from protons and neutrons comprises only a tiny fraction of all matter, perhaps something like 2% to 5%. The world is insidiously dominated by what is invisible, shifting and impalpable.

Open to all influences but without abandoning its light-bound vocation, the venerable science of astronomy is ready to greet dark and non-nuclear matter in order to give meaning to the celestial message.

From the birth of the gods to the advent of dark matter, the visible sky has barely changed. It is not therefore the sky that changes, but our ideas about it. Each new cosmology opens up a new era of human experience. The truly universal matter which governs the future of the Universe, determining whether it is open or closed, may still be unknown to us, but at least we are now convinced of its existence. It is unknown, but not unknowable. Let us venture that by the tenth year of this millennium, it will have revealed its true identity. The discovery of dark matter will be a major scientific event. It remains to wonder whether it will come to us from the sky via astronomy, or from the Earth through the mediation of the particle accelerator.

Astronomy, science of light, has proclaimed the reign of darkness. What will be the consequences of such a grand dialectical turnaround on the intellect and consciousness of humankind? And likewise when the very concept of matter has to be revised? The epistemological programme of the current millennium will involve extending and understanding the concept of the Universe.

In the beginnings of astronomy, the Universe of Aristotle and Ptolemy held only the Solar System. Then it reached out to grasp the Galaxy, and finally the whole system of galaxies. Astronomy is no more than a series of Copernican revolutions, each resulting in a vaster decentralisation and extension of the concept of the Universe.

The Universe has no centre. No privileged place can be taken for the origin of our coordinate systems, for the centre of the world. The last revolution is the most radical of all, for it touches the very stuff of the cosmos, its very materiality. With all the assurance of its well-oiled atomic mechanics, humankind declares: 'Matter that shines and radiates is a mere froth on the surface of dark matter and quintessence.'

But the bell has not yet tolled for anthropocentricity. For we live in an exceptional age when a sample of nuclear matter has managed to express itself through a sentient language, with a voice that rings out, and not just through the blind forces of the microcosm. This tiny, conscious and voluble sample utters a coherent discourse – cosmology – which gives a certain meaning to the word 'universe'.

Our planet has borrowed its matter from the stars. We may carry our heads high: our atoms were nurtured in the stars. We are all of celestial lineage! Human thought is thus brought closer to the immediate reality of sensorial perception. The objects around us, the things we see, the ingredients of our everyday lives accompanying us at all times, the substances we move through and even the association of atoms that make us up all derive their reality from cosmic processes. The Big Bang and the stars are the agents of chemical evolution in a cosmos which today contains speaking matter.

The beautiful exploding flowers we call supernovas illuminate a moving theme indeed: matter's slow ascent to glory. The sky is no empty theatre. Stars are concrete objects like milk and dates. They are the cosmic progenitors of our worldly atoms. Generously they open like blossom, showering space with their myriad winged atoms.

But more than this, thinking matter has endowed itself with a cosmic affiliation. It has bestowed meaning upon its past, composed of inert, stellar and cloudy matter, and before that, the vacuum (although the latter is now considered to be seething with activity in quantum physics).

For the purposes of the present book, we have chosen to speak of atomically luminous, material dust, so precious it seems to us. It is the material frame of beings and structured things, of flesh, birds and stars. Let us not yet provoke the incomprehensible darkness as it lies sleeping in its den. Let us leave in shadow this matter which does not speak the language of light.

From this point onwards we shall take the word 'matter' to have its usual meaning, as the material fabric making up beings and visible or otherwise manifested things. In technical language, this kind of matter is said to be baryonic or nucleonic, that is, essentially made up of protons and neutrons.

Ordinary, everyday matter as we find it in trees and stones, flowers and streams, blood and tears, wine and butterflies, should indeed be qualified as precious, luminous and heavenly, for it is rare and photosensitive and has celestial origins.

Let us retain three simple precepts concerning nuclear matter, to guide us through the following discussion:

- anything hot shines;
- anything that does not shine, which is cold and diffuse, absorbs light;
- anything that emits or removes specific notes of light can be analysed chemically.

Let us bear in mind, so as not to disregard the role played by dark nuclear matter, that anything which neither shines nor absorbs light must nevertheless attract other objects and modify their motion, hence giving itself away.

Light is a conscientious messenger. It carries messages diligently from one point of the atomic universe to another, for it is the very expression of its finer features. Atoms in stars speak the language of the visible to atoms in eyes. No more is needed for a whole astronomy of light and shade, of stars and clouds to blossom forth.

3

Visions

Glossary

absorption line absence of radiation over a very narrow frequency band

blackbody purely thermal emitter of electromagnetic radiation

Boltzmann's constant physical constant relating the average energy of particles to the temperature of the medium they make up

bremsstrahlung radiation due to the rapid deceleration of an electron

CCD (charge-coupled device) electronic detector which has superseded the photographic plate in astronomy

Cepheids stars with variable brightness whose period is related to their luminosity

Compton effect momentum transfer from a photon to an electron

emission line brightening in the spectrum over a very narrow frequency band

frequency of a wave number of wave peaks passing per second

globular clusters groups of ancient stars orbiting around galactic disks

kelvin (K) unit of thermodynamic temperature, with the temperature interval the same as 1 °C; the freezing point of water (0 °C) is 273.15 K

hertz unit of frequency, 1 Hz corresponding to one event per second

interferometry interference of light for observational purposes

inverse Compton effect momentum transfer from an electron to a photon

neutralino hypothetical particle predicted by so-called supersymmetric particle theories

non-thermal radiation emissions arising from the acceleration of charged particles.

Planck's constant (h) immutable number relating particle energy, a corpuscular property, to wavelength, a wave property, in quantum physics

pulsar neutron star with a high magnetic field, emitting narrow beams of radiation, rather like a lighthouse

quasar extremely luminous objects, among the most distant yet observed
redshift stretching of wavelengths induced by the expansion of space in cosmology and denoted z
spectrum distribution of light intensity as a function of wavelength, frequency or energy
synchrotron radiation radiation from electrons guided by magnetic fields
temperature measure of thermal agitation
thermal radiation synonym for blackbody radiation

The language of light: equivalence of colour, wavelength and temperature

Unplaiting the braids of light, we discover the colour blue. Weaving together the strands of blue, red and yellow, all are swamped in white. Light thus looks white if it has a similar spectrum to the Sun, and other colours can be described as a divergence from the latter. However, the word 'colour' is not precise enough to qualify this attribute of light. Of course, scientifically, each note of light, each elementary tone of green, yellow, crimson, mauve or any other hue, is designated a number called wavelength. The tints of light are thus quantified, each assigned its corresponding wavelength. Blue is just a length, expressed by a number (roughly 500 nm), and likewise yellow (around 600 nm) and red (650 nm). So yellow lies between blue and red, and that is all there is to it!

On extragalactic and cosmological scales, light is reddened by the receding motion of its source. The further the source, the faster it appears to move away and the more the source is reddened. However, distance across space also goes with remoteness in time and the past of the Universe is tinged with red. Still further back, it even slips into the infrared.

Light is a conscientious messenger, carrying information from one point of the Universe to another. Atoms in stars speak the language of light to atoms in eyes. Why should we move when light can bring this wealth to us? Such is the resolutely lazy philosophy of astronomy. But do we have the ingenuity required to extract the full panoply of secrets from each passing photon?

Interpellated by the telescope and summoned by the spectrograph to display their luggage, photons from all destinations undergo relentless interrogation. Invited to present their spectrum, they do so without wavering. In their role as customs officials, astronomers examine the data as they might a passport, seeking out the distinctive signature of specific atoms or continuous emissions that could reveal the physical characteristics of the source. The identity sheet might read as follows:

radio □   infrared □   visible □   UV □   X ray □   gamma ray □

continuous □   discontinous □

thermal □   non-thermal □

In this way, all kinds of light, visible or invisible, are catalogued and classified. The range of wavelengths or *spectral band* defines the colour, whilst the terms *discontinuous* or *continuous* specify the presence or absence of spectral lines, i.e. narrow frequency bands featuring higher or lower intensities than the background. The categories *thermal* and *non-thermal* refer to the conditions under which the light was emitted and the physical processes that generated them. Thus, the term 'thermal' relates to a hot gas or plasma source, whilst 'non-thermal' refers to an accelerated particle source.

The visible and the invisible

The human eye is only sensitive to certain tonalities of electromagnetic radiation, in particular, those wavelengths lying between 400 and 800 nm. In fact we are blind to almost the whole spectrum of light. It is of course the perseverance of the Sun's radiation that has fashioned our eye (Fig. 3.1). Acting as a censor to other wavelengths, the heightened sensitivity of the retina to the predominant solar photons explains not only the dark of the night, but all the qualities and failings of our perception.

Today we can no longer ignore the idea that the extreme specialisation of the human eye is a severe handicap when surveying the Universe as a whole. For many stars are much colder or much hotter than our own. Not to mention sources of non-thermal radiation, whose spectra fall a long way outside the frequency bands in which stars tend to radiate. If we wish somehow to perceive this luminous otherness, then we must invent new tools, both mental and experimental, so that we may call ourselves astronomers of the invisible.

In search of such universality, a language of light must be developed which is common to all forms of electromagnetic radiation, whether they are visible or invisible, reducing colour to a reality which is not coloured, expressed in terms of space, time or energy.

The solution is to encode light using numbers. Indeed, it is the way of science to quantify. Our clocks transform time into numbers. Likewise, blue is spirited

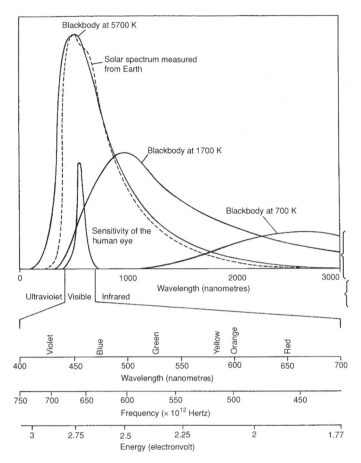

Fig. 3.1. The spectrum of the Sun and the sensitivity of the eye.

away, to become a mere length, as are all the colours we know and love and with them even those that escape human perception, such as X rays and gamma rays. Light is thereby bleached.

In its extended meaning, colour is now a number with units of length, frequency (like our own pulse) or energy. These three physical ideas, spatial, temporal and energetic, are canonically and invariably related. It is a simple matter to translate frequencies into wavelengths or energies, and conversely. Radiation can be treated just as well as a wave with a certain frequency or as particles, photons, with the corresponding energy:

$$\text{energy} = \text{Planck's constant} \times \text{frequency},$$
$$\text{wavelength} \times \text{frequency} = \text{speed of light}.$$

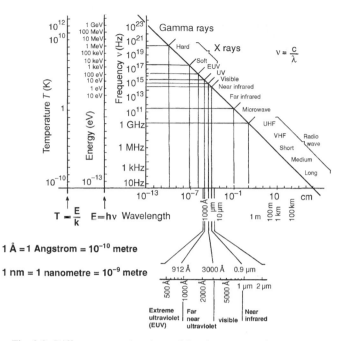

Fig. 3.2. Different spectral regions of the electromagnetic spectrum.

As far as it is merely light, blue has become a number followed by a unit of length. Blue = 500 nanometres. The ultimate physical vector of blueness is a grain of light carrying a few electronvolts of energy. And when its pulse beats 2×10^{15} times per second, light is said to be blue. When this rate falls slightly to, say, 1.5×10^{15} beats per second, the eye–brain detector perceives it as red. Blue has pace. There is punch in the tempo. But then there are the X rays, a form of light with characteristic energy of 10 kiloelectronvolts (keV), and at the other extreme, radio waves pulsating at a mere 10 gigahertz (10 billion vibrations per second) (Fig. 3.2).

Apart from colour, wavelengths evoke another sensation to the physicist: the notion of hot or cold. For tint and temperature are related. When an object is heated at the forge, it changes from black (lack of colour), through red, yellow and white to pale blue.

In fact, blue = 50 000 degrees. Blue means hot, we might say 'blue-hot'. Any object heated to a certain temperature emits radiation known paradoxically as blackbody radiation, regardless of its shape and composition. The wavelength at which this radiation reaches a maximum is inversely proportional to the temperature of the emitting object. The hotter the body, the shorter the emitted wavelengths. The relevant relation between energy and temperature is

$$energy = Boltzmann's\ constant \times temperature.$$

We may thus establish a correspondence between the language of ovens (with waves and microwaves) and the language of particle accelerators. An energy of 1 electronvolt (acquired by an electron in a potential difference of 1 volt) corresponds to about 10 000 degrees. Therefore, an energy of 1 million electronvolts (1 MeV) corresponds to 10 billion degrees. In the aftermath of the Big Bang such energies and temperatures were commonplace. Working side by side thermodynamicists, studying heat, and particle physicists were soon to be joined by cosmologists. For it was clear that, as time went by, the Universe must have cooled as a result of the expansion of space. Heat, energy and time were inextricably linked together.

Universe, tell me how old you are, and I will tell you the colour of your radiation background and the energy of each of your photons. Today, the cosmic background radiation is red, very red. It is so red and cold (about 3 K) that it cannot be seen. Its chilled voice quivers in the great ears of our radiotelescopes. Solar emissions, on the other hand, can be compared with the radiation from an incandescent body at a temperature of around 5700 K. Temperatures vary across the Universe, from 2.73 K for the cosmic background to 100 billion K when a neutron star has just emerged.

Disregarding for a moment the bright and dark bands that decorate the spectrum of a heavenly body at specific wavelengths, the overall hue of that spectrum can tell us the surface temperature of the object. A blue star is thus hotter than a yellow one, and a yellow star is hotter than a red one. The Sun is hotter at the surface than the red star Antares, which in turn stands as a torrid desert before the brown dwarfs or interstellar clouds. The stars go red with cold.

The code used by the plumber for the bath taps is precisely opposite to the astronomer's rule of thumb. Between the artisan and the astronomer, Goethe chose the first. For him, and for all artists since the beginning of time, blue has been associated with what is spiritually cold. In his *Theory of Colours* (*Zur Farbenlehre*, 1810), he wrote that blue expresses a purely empirical psychological impression of cold. Our own science places all sense perceptions such as sound, colour and heat firmly within the human sphere. Nothing outside the human being corresponds to these qualities.

There is another illusion. The blue of a flower, of satin, or of the sea, indeed the blue of any illuminated object, is but deception. The colour of a body lit by the Sun or an electric light is merely the colour of the radiation which it repels. The cornflower in the sunlight is anything but blue! Its atoms swallow all the colours but regurgitate the blue. It might be called 'antiblue'. This is indeed why it appears to be blue. True blue is the blue of heat, emitted and not debased.

Imagined journey in the temperature–luminosity plane

Let us begin by distinguishing the appearance of a star from its essence. Outwardly, the star may be small and white, or it may be huge and red. But beyond these obvious features lie many hidden truths that can only be interpreted through some mathematical and physical model. This theoretical construction brings to light the basic mechanisms at work in the invisible depths of the star, following their evolution through all the layers of its great bulk, right up to the surface that is actually observed.

Stars differ by their splendour and colour, but even more profoundly, by their mass and age.

On the road that leads from appearance to essence, the astronomer's first care is to allow for the effects of distance. To this end all stars are artificially located at the same distance in order to compare their brightnesses in a completely objective way. Curiously, this ploy restores the old idea of a celestial sphere, a purely mental construct so long discredited, dedicating it to the cause of objective knowledge.

Multiplying the apparent luminosity of a heavenly body by the square of its distance, the astronomer calculates its true brightness. The stars can then be sorted, separating out remote bright stars from nearby faint ones which might otherwise appear to be on a par. Colour, on the other hand, does not depend on distance, once corrections have been made for reddening due to interstellar dust.

Let us draw a grid as for a crossword. Up the side we mark the true or absolute luminosity, corrected for distance effects. Along the bottom we put the reciprocal surface temperature (related to colour), with hot on the left and cold on the right. We now plot the stars on our grid. The illustration thereby obtained is one of the most illuminating in astronomy. It is known as the Hertzsprung–Russell diagram (Fig. 3.3). It allows us to pick out a genuine evolutionary path for stars of different masses, depending on the matter being consumed as nuclear fuel within them (hydrogen or helium) (Fig. 3.4). We shall consider this point in great detail in later chapters.

The exact trajectory etched out by each star across this engraving depends on its mass, and to a lesser extent on its richness in metals at birth. Each position along a given evolutionary path corresponds to specific processes of nuclear fusion. For example, the main sequence relates to core hydrogen burning, whilst red giants are engaged in core helium burning. In the asymptotic giant branch, both hydrogen and helium burning take place in shells around the central core.

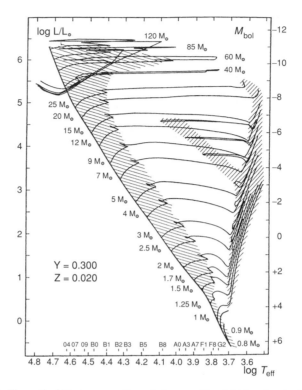

Fig. 3.3. Theoretical Hertzsprung–Russell diagram. The right-hand scale shows in absolute bolometric magnitude what the left-hand scale expresses as the logarithm of the intrinsic luminosity in units of the solar intrinsic luminosity ($L_{\odot} = 4 \times 10^{33}$ erg s^{-1}). On the horizontal axis, the logarithm of the effective temperature, i.e. the temperature of the equivalent blackbody, is put into correspondence with the spectral type of the star, as determined by the observer. This temperature–luminosity diagram shows the lifelines of the stars as strands combed out like hair across the graph. With a suitable interpretation, i.e. viewed through the explanatory machinery of nuclear physics, it opens the way to an understanding of stellar evolution and its twin science of nucleosynthesis. (Courtesy of André Maeder and co-workers.)

Each star follows a different path and at a different rate. Ageing stars turn red, except for the most massive, which become violet or even ultraviolet, gradually moving away from the main sequence. Their core temperatures and pressures increase, thereby triggering further nuclear reactions which can build carbon from helium as the stars ascend the giant branch. The construction of nuclear species in massive stars reaches its apotheosis in the explosion of type II supernovas.

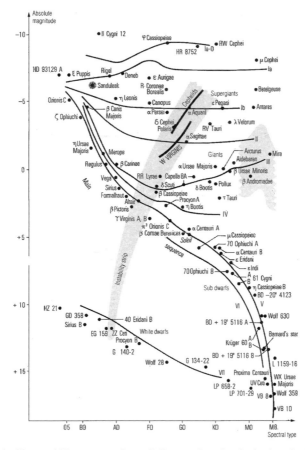

Fig. 3.4. Observed Hertzsprung–Russell diagram. Luminosity is given in absolute magnitude (about +5 for the Sun). The point marked 'Sanduleak' is the progenitor star of the 1987A supernova as it was observed before the cataclysm. On the horizontal axis, the spectral type is given instead of temperature. (After Kaler 1997.)

Notes of light

In 1666, Isaac Newton placed a prism across the path of a light beam in a dark room, and so was born a rainbow. He called the many-coloured iridescence springing from the glass a light spectrum. This dissection of light brought out the manifold colours making up white.

Indeed, any electromagnetic radiation can be decomposed into its different colours and the variation of intensity (i.e. the number of photons) with wavelength (or, equivalently, frequency or energy) is referred to as its spectrum.

Still more generally, the whole range of all visible and invisible forms of light constitutes what is called the electromagnetic spectrum. The word itself comes from the Latin word for 'appearance', like the term 'spectre'.

In the early days observed spectra were simple continuous bands of colour, or chunks of rainbow. However, in 1814, Joseph Fraunhofer was able to magnify the spectrum of sunlight. In doing so he discovered a whole host of dark lines, or absorption lines, where the light intensity was reduced over certain very narrow wavelength ranges. Later, emission spectra were observed by two chemists, Robert Bunsen and Gustav Kirchoff. When light from a chemical flame was passed through a prism, in the very same position as the dark lines in sunlight, they discovered a series of bright lines superposed on the coloured background. Subsequent studies showed that each element possesses its own characteristic series of emission and hence absorption lines.

The wonderful discovery of spectroscopy would allow scientists to set up a correspondence between spectral lines and the chemical identity of emitting or absorbing elements, as well as their relative proportions.

Astrophysicists read such spectra as musicians read their score. When they discern clear notes, they compare them with those emitted by known substances set alight or traversed by sparks in the laboratory. The atom is a violin casting out notes of light. Sometimes they observe an absence of light and match the missing notes with those absorbed by various substances as they are traversed by different kinds of light in the laboratory. They confront the stellar spectrum, bestrewn with absorption lines, with a reference spectrum. The atom is then likened to a sieve that lets past only a certain grain of light. The elements making up the absorbent screen interposed between light source and observer can then be identified, providing analysis of the outer layers of stars and interstellar gas clouds.

This is indeed a scientific revolution: spectral analysis means we may analyse the chemical composition of luminous or absorbent gases without ever even touching them. When asked how it is that astronomers can know the composition of the stars, we may answer that they have learnt to decode the subtle talk of light. Atoms emit or absorb notes of light. This is perhaps the most providential circumstance in astronomy.

Each atom releases or confiscates specific tones of light, depending on its own nature and the external conditions to which it happens to be exposed.

When astronomers see red, the particular red of nebulas, they say: 'hydrogen heated to 10 000 degrees', whilst the blue of nebulas is like that of the sky, for on Earth, the sky is made blue by air (Leonardo da Vinci). Each note of light is associated with a specific atom in a given state. Each note is therefore the sure signature of that atom.

What is more, the shape and shift of spectral lines tell us about the motions of emitting or absorbing atoms. Temporal variations in light emissions inform us of the stable or changing nature of the phenomena that cause them.

One by one, the astrophysicist detaches these notes of light from the spectrum: hydrogen, lithium, carbon, oxygen, iron, barium, europium, the music of the elements. In the spectra of the heavenly bodies, we discover the same atoms that make up our own bodies, and everything else we know down here on Earth. You, the stars, and I, earthling, are made from the same species of atoms. There is unity of substance, but also unity of law. Gravity and electromagnetism, and all other laws of physics, are valid down here and up there, on the Moon, in the Sun and quasars, and throughout the whole observable Universe.

Chemical analysis through light emissions allows us to determine the relative abundances of the various elements in the atmosphere of any star or cloud, and this over the whole range of photon energies. Indeed, spectroscopy has gradually been extended to cover a large part of the electromagnetic spectrum, on either side of the visible. In addition to the originally developed optical spectroscopy, radio, infrared, UV, X and gamma counterparts have been brought to bear on the picture. The detectors are different, but the instrumental strategy and theory remain the same.

To begin with, we must specify the spectral window we are referring to and then use the appropriate detector: telescope, radiotelescope or space-borne observatory. The next characteristic is the accuracy of the energy, frequency or wavelength measurement, followed by the accuracy of the angular measurement (resolving power), and the temporal resolution and sensitivity of the measurement. Finally, we note the direction, date and duration of the observation for each particular celestial object we choose to investigate.

Once the measurement has been made and recorded, interference and background noise must be eliminated so as to exclude all but the relevant part of the signal. It only remains then to interpret the purified signals according to some decoding grid in the form of a physical model of the object (star, cloud, galaxy or the Universe). In this way, after a long process of measurement, analysis, purification and abstraction, a physical meaning can at last be given to the observational data.

The life of an excited atom

Those initiated in the art of spectroscopy soon realised that to each emission line there corresponds an absorption line located at the exact same wavelength. They noticed that, one way or another, atoms can only absorb or emit a well-defined sequence of wavelengths. The amounts of energy which a quantum system like the atom can acquire or relinquish are governed by atomic physics.

The generally accepted model treats the atom as having a centre occupied by a positively charged nucleus with mass but very small volume. This nucleus houses two types of inhabitant: protons, Z in number, and neutrons, N in number. Around the nucleus moves a total of Z negatively charged electrons. Their number is just sufficient to neutralise the charge on the nucleus. These electrons are held in orbit by a force of electrical origin, due to the attraction of each negatively charged electron by the positively charged protons in the nucleus. The orbital motion prevents the electron from crashing into the nucleus, just as it prevents the Earth from falling into the Sun.

To a first approximation, we may assume that the electrons follow elliptical orbits with the nucleus at one focus of the ellipse. This model of the atom thus resembles the Solar System, with the electrons in the role of planets and the nucleus standing in for the Sun.

Although this simile is now known to miss the mark, its historical importance cannot be denied. The centrifugal force victoriously opposes the electrical attraction and conversely, for each electron. A wonderful merry-go-round (dynamical equilibrium) is the result. In this idyllic version, the revolving motion of the electrons would go on forever, if the atom were not subjected to external influences, namely, collisions with other atoms, electrons and photons. Deformed to varying degrees by these impacts, atoms always tend to restore themselves in the most harmonious way, evacuating the excess energy acquired from the collision.

They are said to decay. In doing so, they return to their previous state, corresponding to the most stable possible configuration, towards which all the others also tend. This so-called ground state is characterised by the fact that it has the minimum possible energy amongst all available states. And since it is the lowest in energy, all higher-placed electrons fall into it.

This decay, or return to the ground state, is necessarily accompanied by the rejection of unwanted energy in the form of photons. The excitation energy is therefore restored in the form of radiation. It is thus easy to understand why heat should lead to light. For temperature is a measure of thermal agitation and the speed of atoms. Collisions are all the more common and more violent as the temperature of the gas, and hence the speed of the particles, increases. The result is that atoms are excited and then decay an instant later, emitting light. Hot gases shine. Heat rhymes with light. Hence, any perturbation of the atom's perfect equilibrium triggers an immediate response in the form of a flash of light. For the main part, infrared, visible, UV and X-ray radiation can be traced back to electrons. Only gamma radiation escapes this rule. For it is the light of nuclei.

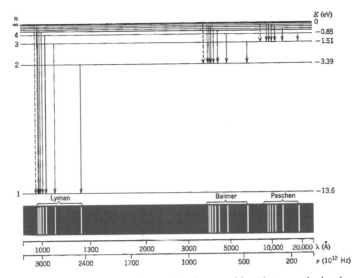

Fig. 3.5. Energy levels of hydrogen, showing transitions between the levels and corresponding spectral emission lines.

The number of satellite electrons dictates the physical and chemical properties of atoms. To a first approximation, they can be considered to be distributed in groups in more or less circular, concentric orbits, located in the same plane.

A more abstract and geometric view is shown in Fig. 3.5. It consists in drawing a vertical line marked off with the possible energy levels of the atom under consideration.

The jump from one level to another then describes a chance excitation and the immediately ensuing decay, with absorption and emission of a photon whose energy equals the difference in energy between the two states. The zero on the energy scale is conventionally attributed to the unperturbed ground state. All the other states, whose energies are necessarily higher, correspond to perturbed or excited states. The first excited state is closest to the ground state, and might be called the first harmonic. States are discretely distributed and only a certain well-established subset of them is accessible to the system. Their energies, however, are not perfectly known because the electron trajectories overlap. In a sense, they follow wide avenues rather than accurately marked paths. For electrons have the vocation to be quantum particles, and 'quantum' is synonymous with 'fluctuation'. They weave back and forth across the whole width of the road as they stagger along.

Fig. 3.6. Synchrotron radiation. A fast-moving electron is forced to spiral along the line of force of a magnetic field and emits non-thermal radiation in the direction of motion of the particle. This type of radiation is confined to a narrow cone whose axis coincides with the direction of motion.

Accelerators in the sky

If red means cold, then infrared stands for colder still. Whilst blue means hot, ultraviolet and X are the signature of even hotter phenomena. But gamma rays are a sign of such high energies that a corresponding temperature would be inconceivable. We thus move from the physics of objects to the physics of beams, from thermal physics to non-thermal physics as it is studied in the great particle accelerators at Saclay in France or CERN in Switzerland, and elsewhere. The range of radiation thereby generated is very different from the spectra of objects shining under the effects of fever, like the Sun and stars.

The distinguishing feature of high-energy astronomy, and especially gamma-ray astronomy, is precisely that it reveals the most energetic and hence the most violent events occurring in the Universe. Gamma astronomy opens our eyes to high-energy processes studied by particle accelerators here on Earth and caused by radioactivity in space.

There is one important exception, however. A certain type of radio wave, called synchrotron radiation (because first discovered in the vicinity of these accelerators), attests to particularly violent phenomena (Fig. 3.6). It is produced in the debris of stellar explosions, the remnants of supernovas.

We now have clear evidence of non-thermal processes in the sky. A whole panoply of violent activities is revealed to the watchful eye of our radio, X-ray and gamma-ray telescopes. Supernova remnants, pulsars, active galactic nuclei and gamma bursts emit radiation that has clearly nothing to do with thermal activity, for their spectra bear no resemblance to those of heated bodies.

It is in this way that we may deduce the existence of natural particle accelerators in the sky, although often concealed. But the greatest and most fabulous accelerator of all is without question the Big Bang.

By a natural process of selection, often brutal, some particles are extracted from their community, in which a certain energy-based democracy normally reigns. Energy is shared in a relatively balanced way across the crowd of particles, although never with perfect fairness. In the centre of the stars, for example, reigns the democracy of heat. Some particles move faster than the rest, but they are rare.

Exceptionally violent phenomena can nevertheless disturb this fine order. Some of them bring tremendous speeds into play, in both translational and rotational motions. Solar and stellar flares come to mind, along with exploding stars and spinning pulsars (neutron stars). If part of this well-ordered energy is conferred upon an atomic nucleus, it then becomes a cosmic ray. It is thus excluded from the community, whose gentle brethren are no longer able to retain it.

An important distinction should be made between electrons and atomic nuclei. The former are easily subjugated and imprisoned by galactic magnetic fields, as a result of their low mass. Spirally through magnetic fields, they emit a characteristic radiation which manifests itself in the radio region of the spectrum. This is the famous synchrotron radiation. Two gamma-ray-generating processes then come into play:

- electrons can transfer energy to the photons they encounter, thereby producing X-ray and gamma-ray photons by the so-called inverse Compton effect;
- electrons passing close by atomic nuclei are accelerated and, like all accelerated charges, generate bremsstrahlung photons.

Launched like missiles, atomic nuclei can smash head-on into other atomic nuclei, fragmenting both projectile and target. Rare and precious species can emerge intact from amongst the debris. Examples are lithium, beryllium and boron, for which nature has found no other means of manufacture.

At still higher energies, let us say 100 MeV and beyond, we move from nuclear physics to the domain of particle physics. In this arena unstable particles are created which release gamma photons as part and parcel of their decay process. When two protons of very high energy run into one another, they sometimes generate neutral pi mesons which then decay to gamma photons.

Viewed through its high-energy gamma radiation, the sky looks very different. Stars are scarce and galaxies hold a pre-eminent position.

The retina and its enlightenment

Seeing is not just a question of adjusting our eyes to the solar spectrum. We live close to a star called the Sun and at night, when it is hidden, we see only stars similar to our daytime star. This does not mean that darkness is absence. The chilled, the scorching and the non-thermal shine invisibly. The eye is in fact doubly solar, for it is made from the same atoms as our star, and it is the persistence and predominance of the Sun's light that have fashioned our sense of sight. The atmosphere is transparent to solar radiation. The maximal sensitivity of our personal detector, the retina, lies in what we choose to call the 'yellow', and yellow is indeed the colour of the Sun.

Furthermore, human beings are daytime creatures, sleeping at night. Their eyes necessarily differ from the eyes of night birds, or butterflies which see in the ultraviolet, off the end of our own visible spectrum. The maximal sensitivity of the retina coincides with the solar spectrum and this is nothing other than the result of adaptation. The constancy of the Sun's emissions is the cause, and this in its turn demands an explanation. What is the secret of the Sun's longevity?

As we shall see in detail, photons (grains of light) produced by nuclear reactions are released at the centre of the Sun in the form of gamma rays, the same lethal rays produced by nuclear bombs. Happily, after wandering for several hundred thousand years within the Sun, softened by encounters with the surrounding electrons, they finally emerge into free space in the form of a constant and gentle light, harmless to our eye.

Each time a proton transforms into a neutron in the Sun's core, a neutrino flies out at the speed of light, or close to it. It whips across the Sun's great bulk in just two seconds, reaching Earth eight minutes later.

In contrast, light staggers through the Sun like a drunkard, only emerging from the great ball of light (the photosphere) after a lengthy perambulation, in a form kind to the eye. During its solar wanderings, the radiation generated in the form of gamma photons by core nuclear reactions transfers a part of its energy to electrons in the medium. (Electromagnetic radiation has a strong affinity for electrons, as the words would suggest.) In this way, the medium is kept hot, and this too is a happy consequence, since heat is crucial for the Sun's survival. Without thermal pressure, it would simply collapse.

Gradually losing their energy, the photons change colour. As they approach the surface, they soften. The ferocious gamma rays give way to X rays, then UV, and finally visible radiation, growing ever softer. Several thousand years go by before the photons accomplish their journey inside the Sun. But then, once released into space, off they fly, straight as arrows, at the astonishing speed of 300 000 km per second which nothing can overtake. If the Sun's core were to switch off at this instant, our descendants would only find out thousands of years from now. Unless, of course, they could detect the neutrinos.

Neutrinos inform us almost instantaneously of what is happening in the Sun's core. However, the main interest of this solar cardiograph is hardly to detect some failure in the Sun's cycle. In capturing solar neutrinos, the aim of contemporary physics is rather to catch the Sun in the act of nuclear transmutation. By measuring the neutrino flux, we may check our understanding of the Sun as a whole and at the same time analyse the relationship between this strange particle and more commonplace forms of matter.

This project has inspired one of the most ambitious experimental programmes of our day. A major contribution has been made by the CEA (Commissariat d'Energie Atomique) in Saclay, France. The neutrino behaves with almost rude indifference towards normal (baryonic) matter. But then we must accept that the matter so familiar to us, making up the stars, the galaxies and our own bodies, is not the dominant form of matter on the scale of the Universe.

The shiny carapace of our daytime star hides a truly enormous nuclear power station, bound by gravity. Each point of its great bulk is attracted within by the huge quantity of matter lying below it. At the same time, it is thrust outwards by the pressure differences that reign across its layers, due to the tremendous thermal activity. Being a gas, the Sun can readjust structurally without explosion. Its temperature and luminosity have therefore remained stable for the last few billion years, making it a wonderful biological incubator and an exceptionally steady lantern.

The fact that our eyes are so exclusively tuned to the Sun has thus blinded us to almost all forms of radiation. This includes radiation from media at very different temperatures, such as the relic cosmological background that filters down to us from the beginning of time, and the great majority of non-thermal emissions, such as the signals from pulsars and supernova remnants.

The eye is thus largely blind to the cosmos, for the solar training it has received can never be completely expunged. Humankind thus suffers from a solar debility. Fortunately, astronomers are able to compensate for the narrow scope of our sense perceptions with the help of electronic prosthetics, artificial retinas placed at the focus of large telescopes or carried aboard space probes. Satellites now bear instruments that greatly transcend human sight. Transgressing the laws of nature, the invisible is thereby revealed to us. We have outgrown the eye given to us by mother nature. Modern telescopes are bigger and better eyes, sensitive to all forms of radiation. Cosmic eyes with universal outlook, open to light of all kinds, these are only human by their construction.

And although our natural and personal detector, the retina, shows us a tranquil sky, with a light scatter of stars across it, striking only by its steadfast inaction, the new sky revealed by telescopes and satellites sensitive to invisible emissions is one of tempest. It is animated by the birth of clouds, the creative explosion of stars and the transition of the Universe from opacity to transparence. Human perception now contemplates regions once forbidden to it.

Light and motion

The spectral study of motion has become one of the key techniques in astronomy. The principle is simple: when a light source (or any source of electromagnetic radiation) is moving relative to the observer, the distance between two successive wave peaks (i.e. the wavelength of the light) measured by the observer differs from that obtained when the source is at rest. For an approaching source, the wavelength is reduced, whilst for a receding source, it is increased. In the first case, we say that the light is blueshifted and in the second, that it is redshifted. This is because the wavelength associated with blue is shorter than the wavelength associated with red.

Furthermore, spectral lines are broadened by random motions of emitting or absorbing atoms. Once the type of atom has been identified, the astronomer can read off its speed by following certain simple rules:

- the width of the lines provides a measure of the turbulent, random and disordered velocities within the system, be they stars, galaxies or clusters of galaxies;
- the shift of the spectral lines indicates the overall (or ordered) motion of the medium as a whole.

This wealth of data is wondrous. Not only do the spectral lines betray the composition of the stars, but they also give away their motions, and even those of the atoms at their surface.

In the case of galaxies, where the light from billions of stars is superposed, their spectra contain the distinct lines emanating from this multitude of sources. The motions of stars to and fro in their midst, and their own motions within clusters of galaxies, both serve to broaden these spectral lines. Moreover, the whole spectrum is shifted towards the blue or the red depending on whether the galaxy is approaching us or moving away. A systematic redshifting of galactic spectral lines, obvious at a glance from the wealth of data available today, tells us the recession speed of the galaxies, either isolated or in groups, and corroborates the hypothesis that the Universe is expanding.

In cosmology, it is more common to speak of redshift than distance or time. The redshift z has an advantage over the time t, insofar as it is a directly measurable quantity. The translation of z into t is model dependent.

To sum up, the all-important art of spectroscopy has shown human-kind that stars are made up of the same atomic species as we find here on Earth, in our own bodies. Furthermore, it has shown us that the Universe is expanding, not the least of astronomy's achievements.

Three-dimensional vision

For centuries, the sole aim of astronomy was to keep up appearances, to calculate the mere countenance of the sky. This wallpaper astronomy remained in essence superficially geometric and two-dimensional. A front view of the Universe was available for observation, whilst its depth remained inaccessible, for want of a distance criterion. The Universe was just a fine cloth hung upon the celestial sphere, splendid indeed, but hardly a true representation of space. One dimension was lacking. The final revelation of the Universe's backdrop, only recently achieved, has been a long adventure.

The distances of stars are today estimated by comparing appearances with reality. High above, an aeroplane seems no bigger than a swallow. We immediately deduce that the plane must be much further away than the bird. Apart from a question of scale, this is the technique used to determine the distance of a star, when it is too far away to apply the well-known parallax method. (The latter was greatly extended by the wonderful HIPPARCOS satellite.) The idea is to compare the light flux of the star with the flux it would have if it were placed at the right distance. Knowing the respective sizes of celestial objects, or else the difference in their natural brightness, we may assess their distance without confusing, for example, bright but remote stars with faint but nearby ones. A sufficient knowledge of the objects combines with observation to situate them at the correct distance.

In this context, certain stars known as Cepheids vary in brightness in a rather regular way. The brightest amongst them fluctuate more slowly. By measuring their rate of variation, we may ascertain the intrinsic luminosity of these pulsing stars. The ratio of the apparent (measured) luminosity to the intrinsic luminosity is proportional to the square of the distance. This provides a way of measuring the distances to certain galaxies, even very remote ones, since Cepheids are very bright objects. One of the missions of the Hubble Space Telescope is precisely to seek out and analyse this type of star in distant galaxies.

The distances to more remote galaxies, however, are obtained by measuring the redshift of their emissions. The redder their light, the further away they must be.

By identifying Cepheid variables in the globular clusters which gravitate around our own Galaxy, Harlow Shapley was able to measure their distance. He thus located their common centre and found it to be a considerable distance from us. It was clear that human beings inhabit the neighbourhood of a nondescript star, very far from the centre of the Milky Way. We are not even at the heart of our own stellar republic! A second assault was thus made on human vanity, after the eviction of the Earth from the centre of the Universe.

Edwin Hubble measured the distance of several Cepheids in the great Andromeda nebula and found that it was situated far beyond the globular clusters in the retinue of the Milky Way. It was then that the milky designation 'galaxy' made its entry into astronomy. The age of extragalactic astronomy had commenced, and it was like leaving the ark. Outside the mists of the Milky Way, and in all directions, uncountable numbers of starry islands could be glimpsed.

The actual distances to the other island universes known today are measured in millions and billions of light-years, whereas human beings only reached out a mere light-second (300 000 km) when they set foot on the Moon. Such a tiny step and we claim to have travelled in space!

Apart from all this, the whole visible Universe with its hundreds of billions of galaxies is engaged upon a uniform expansion. It is not static. The galaxies are moving apart, unless bound together by gravity as in our own Local Group of galaxies, or other clusters.

The Milky Way seems to occupy the centre of the Universe. All around it, the galaxies are moving away at speeds proportional to their distance, as though the Milky Way itself were at the heart of some initiating explosion. But this is all illusion. From any vantage point in a uniformly expanding universe, we would see all other galaxies streaming away, and what is more, observing the same rule that the recession speed would be proportional to the distance. Indeed, only a uniform expansion leads to this situation, in which the world of human beings is not the centre of the Universe. For the Universe has no centre.

The Earth is not at the centre of the Solar System, the Sun is not at the centre of the Galaxy, and the Galaxy is not at the centre of the Universe.

Following this new Copernican revolution, far from being horrified by the remoteness of the stars and by the tremendous divergence operating across the whole Universe, astronomers set about drawing up an overview of the new paradigm and determining its structural and chemical evolution.

History of the world's structure

Astronomy has a distinct advantage over other physical disciplines in that the observer is supplied with a retrospective, a kind of wing-mirror view. Light reaching us from Andromeda is witness to the state of the galaxy 2 million years ago. The huge distances involved and finite speed of light travel allow us to study the composition and behaviour of celestial objects in the distant past. In the case of extragalactic astronomy, we may thus witness galaxy evolution live, as it were. It also provides useful data for investigating the evolution of stellar populations and star formation, not to mention the conditions leading to the birth of galaxies themselves and the emergence of even larger-scale structures. A relatively new field of research has opened up, related to the improved collecting power of telescopes. This story, told by light, is now legible to us.

There are many obstacles to extragalactic astronomy. The main difficulty is that the very great distances involved imply extremely limited sizes and apparent luminosities. In addition, large distances mean spectral shifts from the interesting spectral region (blue) into the near infrared, which is difficult to detect.

For the first time clear observational clues have been obtained concerning the formation of the first star systems. Considerable theoretical progress and much improved numerical simulation techniques have also been brought to bear. New ground-based and space-borne instruments are currently bringing in a vast harvest of data on ever more remote galaxies and supernovas. We find ourselves at a critical stage where the interaction between observation and theory is likely to open up a whole new understanding of the large-scale structure of the Universe.

History of the world's content

By endowing itself with a three-dimensional picture of the cosmos via the distance criterion, astronomy has been able to carry out an in-depth study of the Universe and hence determine something of its architecture. Once described in three dimensions and geometrised on this large scale, matter can deliver up its geographical, or rather its cosmographical, attributes. The structures that appear across the sky are a new kind of constellation: galaxies, clusters of galaxies and sheets of clusters.

This world cannot be motionless. It must move if it is not to collapse. The planets revolve around the stars, whilst the stars whirl around the centre of the Galaxy. Galaxies themselves form moving swarms called clusters, and these clusters move away from one another, swept along by the expansion of the Universe. If they did not, they would simply hurl themselves together under the effect of gravity. The Universe is expanding, and it is this that saves it from collapse. Everywhere and on every scale, motions clearly oppose fall.

But this mechanical world of position and movement does not satisfy the chemist, who immediately asks after its composition. To complete our picture of the world, we must flesh it out with a substance that is no longer an ether, but as concrete as bread or meat. However, this flesh is no simple thing, for it has been worked by generation after generation of stars, genuine factories for the manufacture of atomic nuclei. In addition, its composition varies from galaxy to galaxy, depending on the vitality or indolence of its stellar population. A galaxy that has transformed much of its gas in stars will be richer in heavy elements than a less active one. Only those elements fashioned in the Big Bang, like hydrogen and helium, are distributed uniformly throughout the whole observable cosmos.

Having measured the world's content, the next step is to retrace its past evolution. Time now becomes a parameter. Through the interactions they are involved in, and these depend on the prevailing physical conditions, particles are the driving force for an evolutionary structuring process that implicates time.

The aim of this intellectual discipline which we call astrophysics is to try to understand the unifying features of the inventory of light and matter, visible and invisible, whose character seems so chaotic and unbounded: to discern its physical basis, consistency and wholeness.

Astronomical practice

Astronomy is infinitely subdivided. There are as many astronomies as colours in the visible and invisible spectra, and even more if we include the study of non-electromagnetic signals like neutrinos and gravitational waves.

Notwithstanding, the astonishing diversity of astrophysical instruments should not be allowed to obscure the unity in its method. The complex filters through which we view objects in the sky always deliver the same type of message, in the form of a number. This number is a count rate, that is, a number of photons (grains of light) over a given time, in a given direction and at a given energy (wavelength). Systematically varying the gathering time and the angle and energy of the measurement, we obtain a changing table of data which we may call the observable Universe. This Universe is a table of data.

The main characteristics of celestial bodies are revealed by patient study of images and spectra via a multitude of specifically designed instruments. The information gathered concerns the outward appearance of the objects, such as outline, structure and velocity, but also their more intimate features, including temperature, chemical composition and internal motions. Brought together with the help of a relatively simple and unifying theoretical model, engraved with the exquisite tool of mathematics, these features make up what we hope to be a lifelike portrait of the Universe and its components.

Cosmic archives

Astronomical data is archived and released in the form of images (maps and outlines) and spectra (distribution of photons as a function of their energy). Duly classified, these constitute a huge database. The problem then is to give physical and astrophysical meaning to these cosmic archives, via an interpretation within the framework of the most relevant physical theory.

The quality and insight of the theoretical interpretation or model must relate to the accuracy of the observations. The secret of astrophysics resides in the balance between the refinement of the model and the level of accuracy in the data. Too crude, it is just a caricature; too sophisticated, it becomes absurd.

The men and women of science, mathematicians and physicists, can be divided into two groups. There are those who like to generalise, building up a towering pyramid of knowledge and standing aloft to gaze across their edifice. But there are also those who prefer the details and seek perfection through them. These visionaries and perfectionists, driven by a deep love of science, may belong to the same research team.

Whatever category they may belong to, astrophysicists set out to build, organise and interpret cosmic archives. Their purpose is to give meaning to the messages from the sky. By collecting images and spectra from a vast field of celestial objects (planets, stars, interstellar clouds, galaxies, sky background, etc.), they constitute a cosmic harvest with which to make their bread, good theoretical bread.

Telescopes

Prodded into action by the leaps and bounds of modern technology, the venerable astronomy has rediscovered the enthusiasm of its youth. Today, there is revolution in the air. The exponential growth of astronomical discovery in modern times can be directly correlated with the availability of new detection methods and instrumentation. The science of the sky has become an industry and the great astronomers appear as relics in a landscape completely awash with electronics and data processing.

From Hipparchos to Tycho Brahe, over almost two thousand years, observation instruments remained practically unchanged – the mural quadrant, the triquetrum and the armillary sphere, heavy wooden or even stone devices the size of a man and often fixed like monuments. Glass and metal would revolutionise astronomy, as would the photographic plate and electronics.

Today the contemplative and outdoor science of astronomy, pure and idealistic in its vocation, displays a disconcerting pragmatism through its wide-ranging arsenal. Urania is ready to make the best of any opportunity. She has reaped the benefits of improved navigation techniques (which thus pay back a longstanding debt, for travellers once used the stars to find their way), not to mention the development of communications, computation, sensing, logistics, electronics and ballistics (where rocketry is concerned). Indeed every branch of contemporary technology has been plundered. It follows that discoveries of cosmic significance have been made by a new engineering-style astronomer, not just in the optical region (the Hubble Space Telescope springs to mind), but across its vast complement of invisible rays in the electromagnetic spectrum.

The solitary astronomer with one eye permanently held against the eyepiece of his or her telescope is an image of the past. Today, we see with computer-regulated instruments, using the new techniques of active or adaptive optics.

Astronomers have built themselves huge radio dishes, and space telescopes sensitive to infrared, visible, UV, X and gamma radiation. The latter, based in space, are a kind of extraterrestrial extension of our own bodies, sensorial prosthetics as it were. A satellite-borne detector, ever vigilant, is a roving eye that never closes, seeing the invisible. It is a brain gathering data, but also a tongue, spewing out binary messages and transmitting them over great distances by telemetry. From Earth or space, information flows incessantly and the world is turned upside down: the sky is permanently lit up, shining in the invisible, without respite. The day has no end. The word 'nocturnal' is merely vague and poetic. No, indeed, the night is not dark, it is only our eyes that are obscure. Darkness is not absence, no more than the so-called vacuum.

From the beginning of time, humankind could only admire stars similar to our own Sun, neither much hotter, nor much colder. Non-thermal processes based on high-speed particles, totally escaped our gaze and the main aspects of the sky were hidden to us. But today, the solar eye has given way to a new, universal sense of sight. We no longer live in blindness among the sublime realities of the sky.

One of the great revolutions of the twentieth century was therefore sensory: we extended the range of our sense of sight, thus transgressing a morphological dictate. No vision is now excluded. Humanity has finally removed its blinkers.

So we have discovered that certain scenes of the cosmic show occur in the invisible. The most disturbing is perhaps this flux of chilled photons carrying over from the Big Bang, which comes to us invisible and inoffensive from the very depths of time. Cool astronomies have blossomed in the quest for red, and beyond it, infrared and millimetric radiation, whilst at the other extreme, beyond the violet, new forms of heat have been revealed to us, if indeed the notion of temperature still has meaning beyond a certain limit.

Telescopes in the sky, on the ground and in the basement

The air renders us insensitive to radiation from space, some forms of which are lethal, rather like a skin-protecting cream. At the same time it acts as a censor with regard to the astronomical information carried by that radiation. There is no choice therefore but to break out from the cocoon, to rise above the atmosphere by means of stratospheric balloons, rockets and satellites. With its airborne and space-borne telescopes, the whole planet Earth is turned towards the Universe, its eye emerging from the air like a bather from the sea. However, the sky is not the only arena of modern astronomy. Telescopes are also flourishing in the basement and continue to prosper on high mountain tops.

Running in parallel to spectacular space-based astronomy, optical telescopes and radiotelescopes have progressed in a quite breathtaking manner thanks to the new techniques of interferometry and active and adaptive optics. Telescopes perched on mountain peaks, such as the CFHT (Canada–France–Hawaii Telescope) and Keck on Mauna Kea in Hawaii and the VLT on Cerro Paranal in Chile, and radiotelescopes set out like windmills in Puerto Rico, Sologne (in the French Alps), The Netherlands and Spain, gather photons able to cross the layers of the atmosphere without major alteration, whilst spectrographs then dissect the radiation into its finest detail.

Neutrinos have inspired a very special kind of astronomy, since detectors are constructed underground. A new species of cave-dwelling astronomer waits down a disused gold mine or other such subterranean retreat in the hope that a handful of these elusive particles will fall into the fabulous traps they have laid out. From time to time, one among the billions of passing neutrinos will be ensnared in an enormous chlorine, gallium or water detector. For the neutrinos pass through rock as we move through air.

These same physicist astronomers, today called astroparticle physicists, are currently setting up and fine-tuning traps for neutralinos, hypothesised particles of dark matter. These are just as subtle as neutrinos, but much rarer and much more massive. For the moment, the search has been fruitless, but patience is a virtue in the hunt for the invisible.

Another new style of astronomer is interested in gravity in its most fundamental form. They are at present working on huge interferometers which should detect gravitational waves, induced by distortions and deformations in the elastic fabric of space–time itself, and even harder to lay a hand on than the neutrinos and neutralinos. Still other astronomers are working on high-precision satellites to investigate the finer predictions of general relativity, a huge project with many partners, including the French space agency CNES (Centre National d'Etudes Spatiales).

Astronomy of the senses: images and spectra

Depending on the type of detector pointed at the sky and the wavelength region to which it is sensitive, the objects revealed may exhibit different levels of energy and violence.

1. Cold and soft: radio, microwave and infrared radiation. Telescopes
   sensitive to the gentle radiation of the infrared unveil some of the more
   tender scenes in space, such as cloud formation and the birth of stars. They

also show us the ancient and icy cosmological background which carries in its anisotropies the imprint of all future galaxies in their larval state, like a watermark; or again, the glittering clouds between the stars, full of dark chemistry, and dwarf or aborted stars that shine coldly.

2. Temperate and calm: the visible spectrum. For the main part, visible stars are passing through a mature and gentle phase. Like the Sun, most lie on the main sequence of the HR diagram, slowly burning hydrogen in their core. Healthy stars lead a balanced lifestyle, with stable, long-lasting, visible emissions, like our own Sun. The optical sky is a picture of calm and stability. Indeed, it gives a misleading impression of peaceful tranquility. From time to time, it does happen that a 'new' star may come to disturb this glassy surface, for a few moments. But in reality, it is often just the last cry of a star that is leaving the worthy company. Star explosions show up by the intense light they release in the visible.

3. Scorching and violent: UV and X radiation. The UV is the ideal spectral region for viewing astromical objects that have been heated to several tens of thousands of degrees. Curiously enough, intense UV sources include both the very hottest and the very coldest stars. The bright emissions of hot stars seem perfectly natural, since hot objects radiate at short wavelengths. We see not only massive and hot stars, but also white dwarfs, which died blanched and beautiful, lustrous remains of past suns, and cold stars endowed with sufficiently hot coronas and chromospheres to manifest themselves through UV emissions, rather like our own Sun.

Some of the more common chemical elements such as carbon and oxygen emit readily in this range of the spectrum. In this way, the Milky Way informs us of the content of its stars and gases.

Still shorter wavelengths produced in even hotter media are more penetrating, to such an extent that X-ray astronomers may claim to see the whole galaxy. In fact, the X-ray galaxy bears little resemblance to the visible galaxy. The main sources are quite monstrous pairs of stars. One normal star, in the main sequence or red giant branch, is torn apart by a neutron star, a genuine stellar corpse, which gradually gobbles down its fleshy partner. Before swallowing the next meal, it heats it up until it gives off X rays. Certain binary systems containing a white dwarf also emit in this wavelength range, but in a more modest way.

A diffuse X-ray background is perceived between these brilliant sources. It comes for the greater part from remote regions outside our Galaxy, probably quasars and other active galaxies. Added to these more distant emissions is a galactic component emitted by the hotter interstellar gases, in particular, those of the local superbubble, a hot enclave containing the Sun and its planets.

The brightest X-ray nebulas are the remnants of supernova explosions. The ejecta are thrown out so forcefully that the collision with neighbouring interstellar gases produces temperatures of several million degrees. This is sufficient to emit photons in the keV range. It is no surprise that most X-ray binaries and supernova remnants should be located in the galactic disk.

Astronomy's new eyes carry code names like HST (Hubble Space Telescope), XMM (X-ray Multi-Mirror) and GRO (Gamma-Ray Observatory). Each wavelength region has its own squadron, extended by future spacecraft that take us ever further towards sensitivity and detailed analysis of images and spectra.

Let us begin with optical astronomy, perhaps the most familiar to us. The big question today is: should it be developed in space or on the ground? Or is there scope for a complementary development in both arenas? When it comes to removing the perturbing effects of the atmosphere, astronomers can be divided into two positions. There are those advocating an Earth-based campaign, whose weapons include interferometry and adaptive optics. And there are those in favour of a space-borne strategy who, in connection with the world's space agencies, fearlessly propose to launch the very telescopes that seem best suited to a firm anchorage on the ground, the radiotelescopes, promoting the development of a genuine space-based interferometry. The discussion centres around the question of value for money. Some rudiments of observational astronomy are needed to follow the debate.

The key word in astronomy today is spectro-imaging. The combination of spectrum and image allows astronomers to deduce valuable information about objects under investigation, such as chemical composition, temperature, internal constitution and motion, or even the distance to the object.

Imaging a source means discerning its fine spatial structure or low intensity details that could not be detected without high-resolution telescopes. For example, it may consist in resolving the stars in another galaxy in order to study their individual features.

Producing a spectrum means identifying as many emission and absorption lines as possible in the light signal in order to extract, in particular, the chemical composition of the emitting object.

Having defined the aim, let us return to the question of observing equipment. The purpose of the astronomical telescope is to gather light from celestial bodies and to focus it. This light is then dissected by means of a dispersive tool such as a prism or diffraction grating placed at the focus of the telescope. Signals are recorded by extremely sensitive electronic detectors called CCDs (charge-coupled devices).

The quality of a telescope is gauged by its performance relative to three criteria:

- sensitivity – ability to see faint objects;
- spatial resolution – ability to distinguish objects at small angles of separation;
- spectral resolution – ability to distinguish spectral features at small wavelength separation.

Sensitivity is directly related to gathering power. That is, considering the telescope as a kind of giant photon funnel, sensitivity is proportional to the area of its mirrors, which focus the light.

In the absence of atmospheric effects, the theoretical spatial resolution of a telescope is inversely proportional to its diameter and proportional to the wavelength of the radiation. An 8-m telescope, such as one of the four components of the VLT, has resolution 0.014 seconds of arc in yellow light, but only 1.3 minutes of arc at wavelength 3 mm. This explains why radioastronomy requires such enormous collecting areas.

Deviations of the collecting surface from an ideal parabola must not exceed a tiny fraction of the shortest wavelength to be analysed. Grinding and polishing are the bread and butter of optical astronomy.

When telescopes are connected together in interferometric mode, the resolution of a network is given by the wavelength of observation divided by the baseline, the latter being the maximal distance separating two collectors. By cleverly coupling these collecting units, it thus becomes possible to emulate a single giant instrument whose diameter would be equal to the maximal separation of two components. However, the sensitivity of the interferometer remains inferior to that of the giant simulated telescope.

Developments have been so phenomenal in the field of ground-based astronomy that this alone has induced a genuine transformation both in the techniques used and in the way of thinking. The ESO (European Southern Observatory) has been the driving force in developing new technologies known as active and adaptive optics that have brought ground-based telescopes back into the league of their space-borne counterparts.

Active optics involves adjusting the mirrors in real time in such a way that they maintain their optimal position and shape at any moment, independently of the direction in which the telescope is pointing. If the image deteriorates in any way, correcting signals are sent to the computer-controlled mirror and it immediately readjusts the optical elements by means of electromechanical active supports.

Adaptive optics is designed to correct distortion produced when the light from a celestial body encounters turbulence in the Earth's atmosphere. A correcting mirror placed on the light path continually changes shape in such a way as to wipe out the effects of atmospheric perturbation. The quality of the image becomes almost as good as if the telescope were in space. In parallel with the development of telescopes themselves, detection instrumentation has also leapt ahead. Today, the best detectors attain almost 100% efficiency in transforming collected light into electrical signals.

Given their complexity and cost, the main projects require huge teams of people. This marks an important sociological change. From this point of view, astronomy has come into line with particle physics, which enlists armies of research scientists and technicians to realise projects of gargantuan proportions. Until quite recently, technical effort focussed largely on construction, but today, equal emphasis is put on the mode of operation of the very large telescopes, as well as their scientific and technical management.

We are now in a position to sketch a brief recent history of astronomy, concentrating on the mainly spectroscopic concerns and highlighting the very real dialectic that has sprung up between Earth- and space-based astronomy, between Europe and the rest of the world.

Large monolithic mirrors continue to catch attention. At the same time, other solutions have been envisaged. Although it has been reserved until recently for longer wavelengths, interferometry is now applied to visible light. Active (i.e. self-correcting) and even adaptive optical systems have been set up on all existing telescopes. Very often, these installations have seen improved image quality and control systems. New approaches to extremely large telescopes are currently taking shape. Progress is impressive and the images obtained quite dazzling (see for example the splendid collection of images of galaxies and nebulas produced by the HST and the VLT).

Optical telescopes with 4 m diameter which had blossomed in many of the world's deserts have now been superseded by the space telescope and 8-m or 10-m telescopes on dry mountain peaks. These giants with their penetrating eye and minuscule field of view are mainly devoted to spectroscopy.

The VLT, centrepiece of the ESO, promises an exceptional harvest of astronomical data. Understanding the chemical evolution of the Universe requires a coordinated study of the most remote objects, ancient stars in the galactic halo and absorbent clouds in the line of sight of quasars. To this end, the high-resolution spectrograph UVES (UltraViolet Echelle Spectrograph) was set at one focus of Kueyen, one of the four components of the VLT, perched at the top of Cerro Paranal in Chile. Kueyen–UVES is the best spectrograph–telescope combination of its generation. It will be principally devoted to determining the abundance of a range of critical elements in the atmosphere of stars in our own Galaxy and nearby galaxies. It will also be possible to conduct a minitomography of the absorbent interstellar and intergalactic media out to a redshift of $z = 2.1$.

American astronomy holds several trump cards. Until recently, its great optical telescopes, the HST and the twin telescopes Keck I and II set up on Mauna Kea in Hawaii, reigned imperially over the visible and near UV sky. Although the European VLT has largely counterbalanced this domination, the USA is ready to step into the lead once more.

Quite apart from its Gemini telescopes representing state-of-the-art ground-based astronomy, the USA is preparing the formidable NGST (New Generation Space Telescope). With a satellite-borne mirror 6 m to 9 m in diameter, it combines a large gathering power with the advantages of viewing from outside the Earth's atmosphere. It will be an excellent tool for viewing in the near infrared for its optics and detector systems will be cooled. However, in order to maintain certain parts of the instrument at low temperatures, it will have to follow a particularly high orbit. This excludes any possibility of in-flight repairs or maintenance by astronauts brought out aboard the space shuttle, a considerable risk when we recall that the HST owes a great deal of its acuity to such post-launch interventions. However, the stakes are high, since this descent towards the red end of the spectrum will allow the new telescope to probe the Universe more deeply than its predecessor. Indeed, it will be able to study with great accuracy the redshifted light of remote extragalactic sources.

As its name suggests, the NGST is indeed designed to supersede the HST. The new space telescope will go into service in 2003. The US space agency has been pursuing feasibility studies since 1996 and this project is a key element in NASA's ambitious long-term Origins programme, counterpart to the European Space Agency's own Horizon 2000 and Horizon 2000-plus programmes.

Optimised for the 1–5 μm wavelength range with extensions into the visible (0.5 μm) and infrared (30 μm), this great winged giant should restore supremacy to the astronomy of the new continent. It is wise to accept the fact and encourage transatlantic cooperation.

NASA has formally invited its European counterpart to work on a continuation of the HST and the ESA has accepted. This in no way compromises its own cryostatic project, FIRST (the Far InfraRed Space Telescope). Designed to supersede the highly successful ISO (Infrared Space Observatory), this new space observatory will probe the Universe in the far infrared and submillimetre ranges. It should be recalled that the ESA contributed some 15% of the costs of the HST, something which is often forgotten. For its part, NASA will be involved with FIRST, thus perpetuating cross-Atlantic ties in astronomy.

The far infrared and submillimetre ranges cover the brighter emission lines of the interstellar medium. This should allow detailed studies of its composition, both in the Milky Way and in other galaxies.

The observational astrochemistry of gas and dust will become an important tool for understanding these cold environments and through them, the stellar/interstellar cycle. It will also help us to understand how stars form and evolve. Almost all stages in the transformation of gases into stars will be accessible to analysis by FIRST: the appearance of dense globules in interstellar clouds, cooling and freezing, formation of disks around protostars, coagulation of dusts and formation of planetesimals.[1]

In the longer term, ALMA (the Atacama Large Millimetre Array) will take over from FIRST. This will be a huge network of 96 radio dishes extending over an area of 10 000 m^2. Its detectors will cover the frequency band from 70 to 950 GHz. Sponsored by Europe, the United States and Japan, ALMA will be built at an altitude of 5000 m on the Atacama plateau in Chile.

Beyond the violet, France is present in all areas of astrophysics, from UV through X to gamma rays. Just mentioning the main items devoted to spectroscopy, France is involved in the satellites FUSE, XMM and INTEGRAL.

FUSE (Far Ultraviolet Space Explorer) combs the sky in search of the spectra of hot bright background objects (quasars and stars) modified by absorption lines due to interposed clouds (both extragalactic and interstellar). Its main prey is deuterium, an atom of major cosmological significance. It should also establish a detailed breakdown of interstellar gases. FUSE is a NASA project with French participation.[2]

Newton–XMM is an X-ray telescope equipped with a set of nested mirrors designed to focus grazing-incidence X rays, a configuration which explains the name X-ray Multi-Mirror. It is an ESA project and was launched by Ariane 5 in December 1999. It opens a window onto the ultrahigh temperature Universe with its explosions and stars ripped apart by black holes. Its spectroscopic targets are supernova remnants and the gases that fill clusters of galaxies.

In the hot X-ray region, the American observatory Chandra–AXAF will be a serious rival for XMM. It has better spatial resolution due to its exceptionally smooth mirrors, but lower sensitivity because of a lower collecting area.[3] Japanese astronomy is also very present at these wavelengths.

---

[1] Laurent Vigroux, the head of the astrophysics department at the CEA, is one of the most active proponents of the FIRST project.

[2] Alfred Vidal-Madjar of the Paris Astrophysics Institute is one of the main protagonists.

[3] Monique Arnaud of the CEA in France is the lynchpin of XMM operations in France.

Curie–INTEGRAL will reveal some of the secrets of nuclear and radioactive processes in the Universe, carrying on from NASA's worthy CGRO (Compton Gamma-Ray Observatory). The new observatory, measuring 9 m in length and equipped with gamma-ray imaging and spectrometry instruments, was placed in orbit by a Russian Proton rocket at the end of 2002. It will scour the sky for nuclear gamma-ray lines, clues to the existence of radioactivity and hence recent nucleosynthesis. It will also map radioactive sources across the Galaxy and, when the opportunity arises, it will analyse gamma emissions from supernovas and novas.

As gamma rays pay no heed to lenses and mirrors, INTEGRAL is a rather special telescope, resembling rather a nuclear physics installation than an astronomical instrument. As a result, its resolving power will be much lower than that of an optical, UV or X-ray telescope. The identification of gamma sources will thus require support from more conventional forms of astronomy. INTEGRAL is an ESA project.

A new era of European supremacy will open up in the domain of nuclear gamma astronomy, for no large-scale project has been proposed across the Atlantic to take up where the CGRO observatory left off. Let us hope that European science will make the most of this boon.[4]

Universal sight

UV, X-ray and gamma-ray observatories are all space-borne because these wavelengths are blocked by the Earth's atmosphere. And what could be more natural than to place eyes in orbits! But who sent them up there and why? Astrophysicists are no longer naive when it comes to the relationship between the tools of military surveillance and those of astronomy. And this is not a recent invention. Galileo himself offered his refracting telescope to the delighted dignitaries of the Venetian senate in order to draw attention to its military potential. His salary was immediately doubled and he was appointed professor at the University of Padua!

The science of the sky has been transformed beyond recognition by technological and scientific progress, particularly when it comes to space-based astronomy. With our hypertrophied sense of sight, in the form of myriad electronic extensions, the astronomer is able to explore a deep and real sky.

Our solar eye has given way to universal sight. This art no longer consists of one colour, but of a combination of all colours, be they visible (red,

---

[4] Technical summaries of the various space observatories are generally available on the Internet. See Appendix 8 for addresses.

yellow, green, blue, violet) or invisible (radio, microwave, infrared, UV, X ray and gamma ray). Between radio and gamma wavelengths, a whole bestiary of luminous phenomena attests to the varying relationship between atoms from one place to another, at different epochs across the Universe. The new astronomer holds out a mirror which reflects all the gentleness or violence of the world.

The choice of visible or invisible colours, i.e. the range of wavelengths, in which an object or class of objects will be observed, is carefully premeditated. Pointing an infrared telescope towards an interstellar cloud, seeking out this gentle radiation, so red that it cannot be seen, the astronomer becomes sensitive to star birth, or emissions from newborn stars letting out their first cry of light from a dusty and cloudy placenta.

Light sifted by dust takes on an infrared tinge. The ancient chilled background radiation of the Universe, the glittering clouds that float between the stars, plying such complex chemistry in their dark confines, cold stars, dwarf or aborted, that shine frostily: all these things can be spied upon by their infrared or millimetre radiation.

For each star that shines, there are a certain number that remain hidden, masked by a veil of dust. Observing the Universe in submillimetre waves, where the heated dust shines at its brightest, it has been possible to show that in the young Universe, stars were born at a rate about five times greater than was suggested by observations in the visible.

The millimetre (microwave) radiation usually referred to as the cosmic background radiation (CBR) was scrutinised to great effect by the COBE (Cosmic Background Explorer) satellite. This background is witness to the state of matter some 300 000 years after the Big Bang. Its characteristic temperature today is 2.735 K on the absolute temperature scale. This perfectly thermal radiation was released when the Universe ceased to be opaque to its own light emissions, just at the moment when matter decoupled from light, thus allowing structure to develop. This was very likely the time when the future architecture of the cosmos was decided. The folds of the cosmic drapery wrapped around the Big Bang suggest that the shape of the galaxies arose from an undifferentiated amalgam of matter and light. Generations of astronomers have been spurred on by the desire to understand how galaxies formed and evolved. It may be that this quest will soon be accomplished.

Armed with a gamma telescope, to pick out this emissary of violence, the astronomer can show us the regenerating explosions that shake the cosmos. Each object has its eye, each eye its object. The refracting telescope reached out to the Moon, whilst the gamma telescope beholds the explosion of stars.

A continuous flow of data falls from the sky like a fountain, into the open mouths of satellites like FUSE, Chandra and XMM. An enigmatic portrait of cosmic violence is being painted in our understanding of nature, peopled by supernovas, neutron stars, black holes and stellar corpses, telling of death and the life beyond it. For the panchromatic astronomy of the invisible, the sky is literally exploding upon our understanding.

4

Contents of the sky: atomic sources and fountains

Nothing exists except atoms and empty space. Everything else is opinion.

Democritus

Glossary

abundance quantity of a given element relative to the quantity of hydrogen

astration star formation

asymptotic giant branch region of the HR diagram in which red giants reach the end of their evolution

baryon heavy particle, such as a proton or neutron

carbon fusion production of neon, magnesium and alumninium

chondrule meteorite inclusion

electron light, negatively charged particles revolving around atomic nuclei

helium fusion conversion of helium to carbon, oxygen, etc

hydrogen fusion nuclear conversion of hydrogen to helium

isotope particular atomic nucleus defined by the number N of neutrons it contains

metallicity quantity of elements with atomic number higher than helium in one gram of the relevant sample

oxygen fusion production of nuclei between oxygen and silicon

photon particle of light

photosphere region around a star from which light is finally released

p process production of rare, proton-rich species

proton electrically positive particle making up atomic nuclei

protosolar cloud interstellar cloud from which the Sun was born

r process production of nuclei above iron involving rapid neutron capture

silicon fusion production of nuclei between silicon and zinc

spallation shattering of nuclei under the effect of violent collisions
s process production of nuclei above iron involving slow neutron capture

Sources of the elements

In the eighteenth and nineteenth centuries, chemists had so successfully isolated the elements that John Dalton was able to put together a genuine atomic theory. Dmitri Mendeleyev organised the elements into his periodic table, the culmination of scientific elegance.

Confirming the idea of atomic structure, J.J. Thomson discovered the electron and Ernest Rutherford the atomic nucleus. When the nucleus was broken, Rutherford succeeded in distinguishing the proton, whilst James Chadwick identified the neutron. J.J. Thomson and his son G.P. Thomson each received the Nobel prize, the first for showing that the electron is a particle and the second for showing that it sometimes behaves as a wave. The discovery in France of natural radioactivity opened the way to the prospect of spontaneous transmutation. In doing so, it also demonstrated that the elements of the periodic table are not eternal. Pauli hypothesised the existence of the neutrino and it was discovered twenty years later. In the meantime, quantum physics had taken a firm hold of the atom and its nucleus, which became its favoured plaything. Nothing relating to the microcosmos escaped its notice or evaded explanation by the new theory.

Today, physical chemistry has accomplished its great task of elucidating the microcosmos. The existence, properties and combinatory rules for atoms have been firmly established. The problem now is to work out where they came from. Their source clearly lies outside the Earth, for spontaneous (cold) fusion does not occur on our planet, whereas radioactive transmutation (breakup or decay), e.g. the decay of uranium to lead, is well known to nuclear geologists. The task of nuclear astrophysics is to determine where and how each species of atomic nucleus (or isotope) is produced beyond the confines of the Earth.

This raises the burning question: starting out from a simple substance (not to say elementary) made up of photons, electrons, neutrinos, neutrons and protons, what mechanisms exist for synthesising the many and varied nuclei to be found in nature? This in turn raises the question: where and when did these processes take place, and how do they fit together chronologically as the Universe has evolved?

Working with this quintet of particles, symbolically denoted γ, e, ν, n and p, nature built up all the elements in the periodic table, conferring their distinctive chemical properties upon them. They then formed simple molecules like water (H_2O) and carbon monoxide (CO), and more complex carbon-bearing molecules that would eventually allow life to appear on Earth. In the end, DNA is just a particularly fortunate arrangement of electrons, protons and neutrons.

Spectral analysis shows quite clearly that the various types of atoms are exactly the same on Earth as in the sky, in my own hand or in the hand of Orion. Stars are material objects, in the baryonic sense of the term. All astrophysical objects, apart from a noteworthy fraction of the dark-matter haloes, all stars and gaseous clouds are undoubtedly composed of atoms. However, the relative proportions of these atoms vary from one place to another. The term 'abundance' is traditionally used to describe the quantity of a particular element relative to the quantity of hydrogen. Apart from this purely astronomical definition, the global criterion of metallicity has been defined with a view to chemical differentiation of various media. Astronomers abuse the term 'metal' by applying it to all elements heavier than helium. They reserve the letter Z for the mass fraction of elements above helium in a given sample, i.e. the percentage of 'metals' by mass contained in 1 g of the matter under consideration. (Note that the same symbol is used for the atomic number, i.e. the number of protons in the nucleus. The context should distinguish which is intended.)

In order to determine which processes nature uses to produce the various atoms, it was first necessary to establish the overall chemical composition of our own reference star, the Sun, moving on then to the other stars in order to classify them within the stellar hierarchy of masses and ages. It is indeed edifying to compare the compositions of various categories of stars of different ages. Each star reveals its composition at birth etched onto its luminous outer layers, and this is the very constitution of the interstellar medium from which the star originated. By this method, groups, populations and companies of stars begin to stand apart. Fruitful comparisons between astronomical objects (viz. the stars) of similar nature but different generations are then possible. This in turn should allow us to elaborate an evolutionary theory of truly galactic, or even cosmic proportions.

The Galaxy

The chemical evolution of the Cosmos is by now an undeniable reality, clearly demonstrated on a more modest scale by careful studies of our own Galaxy. Indeed, considering the Milky Way and its retinue of nearby galaxies, different populations arise that can be classified as metal-rich, metal-poor or average.

Having calculated the age of each group of stars, we may read off the indicators for the chemical evolution of galactic gases as a whole.[1]

A significant part of the metal-poor stars are huddled together in the globular clusters, a kind of gathering of old stars that formed right at the beginning of our Galaxy's history. These dense congregations of stars revolve around the bright disk of our Galaxy. The ages of globular cluster members can be determined from their positions on the temperature–luminosity diagram (the Hertzsprung–Russell diagram discussed earlier). In fact the most elderly among them are at least 12 to 14 billion years old. They have never left their place of birth, held together as they are by the bonds of gravity. These old ladies still shine with a certain lustre. When their light is analysed, we find that they are not only old, but have little personal wealth. Indeed, they reveal a considerable deficiency in metals. The poorest of the poor among them possess less than one ten-thousandth of the Sun's metallic fortune.

They carry the stamp of the Big Bang, whence their great interest for cosmology. Their lithium content in particular is a precious clue as to the nucleonic density of the Universe, combined with deuterium and helium abundances measured in extremely metal-poor media (see Appendix 1).

The first major observation is thus that the Sun is a rich star compared with the ancient stars in the galactic halo, placed like a crown around the Milky Way, but that its composition is close to that of the stars in the disk where it itself resides. Our daytime star therefore belongs to the wealthy fellowship of the disk. Whereas the halo is almost completely devoid of gases, the disk abounds in them.

The metal deficiency of the oldest stars clearly suggests that, through generation after generation of stars, the Galaxy has continually built up its hoard of heavy elements, those so crucial for the formation of life. Hence the accumulation of metals in the Galaxy, as in every other galaxy, is a gradual process. Here then is the theme that we shall promote and develop throughout this book, that it is indeed the stars that have toiled to improve the rich variety of elements present in the Galaxy today.

The Sun

Coming back to our worthy Sun, the relative proportions of the various chemical elements in its atmosphere are determined by analysing the spectrum of its

---

[1] French astronomers, and in particular, Roger Cayrel, François Spite and Monique Spite at the Paris-Meudon Observatory, are world authorities on determining the composition of ancient halo stars.

photosphere, a spherical envelope from which visible light is finally emitted by the Sun. The composition of its inner regions, in particular the core, in a turmoil of nuclear transmutation, cannot be measured directly. It is only accessible by calculation.

According to models of the internal structure of the Sun to be considered later, we may consider that the observable outer layers are in no way contaminated by nuclear reactions, the latter being confined to the central region. Only the extraordinarily fragile lithium seems to be modified there. It is carried to the surface by convection currents in the tremendous agitation of matter, from depths where the temperature is high enough to cause significant degradation. But apart from this exception, we may consider the composition of the Sun's luminous outer region to be the same as that of the cloud from which it formed some 4.6 billion years ago. Like the meteorites, its surface thus comprises the same mix as the protosolar cloud that gave birth to the Sun, the Earth and the planets.

By its great mass, the Sun constitutes the major part of the Solar System. In this sense, it is more representative than the planets, which have been the scene of intensive chemical fractionation. The composition of the solar photosphere can thus be compared with the contents of meteorites, stones that fall from the sky, a second source of information on the composition of the protosolar cloud, provided that volatile elements such as hydrogen, helium, carbon, nitrogen, oxygen and neon are excluded. Indeed, the latter cannot be gravitationally bound to such small masses as meteorites and tend to escape into space over the long period since their formation.

The carbonaceous chondrites, which constitute a tiny proportion of the matter within the Solar System, do conserve within them the original composition of the Solar System. If we exclude the volatile elements mentioned above, these rare meteorites have hardly been affected by the subsequent metamorphism of our planetary system.

Agreement between the two sources of information is excellent. Moreover, laboratory analysis of meteorites can determine the isotopic composition of matter making up the Solar System, an invaluable piece of data for those intent on understanding the origin and evolution of atomic nuclei.

We thus arrive at the following composition for the ancestral cloud that spawned the Solar System: in 1 gram of matter, we find 0.72 g of hydrogen, 0.26 g of helium and 0.02 g of heavier elements. Despite the superb efforts of past generations of stars, the Sun, like its nebulous father, is singularly poor in metals, since these make up a mere 2% mass fraction of its matter. This, however, is a small fortune compared with the ancient stars in the galactic halo.

The Sun does not contain more than 2% of heavy elements (by mass). This meagre total has nevertheless sufficed to engender and perpetuate life and consciousness, as we may deduce from our own existence and the composition of our star. But since the birth of the Sun 4.6 billion years ago, the stars have not laid down their tools. What will life and consciousness be like when they reach 3%, or 10%?

In truth, the question does not even arise. By reference to the iron levels in stars of different ages across the galactic disk, it seems that the metal content has reached a ceiling. We cannot therefore expect any great increase in metals and minerals in the future interstellar medium, and this for a very simple reason: the gas from which the stars form is gradually being used up (see Appendix 5). In our own galactic neighbourhood, it comprises only about 10% of the mass of the disk. All the rest is in stars. The Galaxy seems to be somewhat out of breath.

Chemical evolution is thus coming to an end, even if here and there we may observe a dazzling flash, reminiscent of the old glory. Indicators of recent nucleosynthesis have been provided by observations of radioactivity in the galactic disk. Detection of the gamma-ray line at energy 1809 keV emitted when aluminium-26 decays in different directions has made it possible to draw up a map of the galaxy. This is a radioactive nuclide with a decay time of around 1 million years, no more than a split second compared with the age of the galactic disk (some 10 billion years). In this sense then, we have been able to catch nucleosynthesis in the act, as it were, studying with great accuracy the mechanisms that lead to the formation of aluminium-26. The challenge was to understand how this isotope could be produced by stars and ejected into the interstellar medium before it decays, that is, in under 1 million years. It would appear that its main source is in the supermassive Wolf–Rayet stars which blast material away in a tremendous stellar wind, and also in supernovas, the apotheosis of the stellar condition. Everywhere we look, we find the same pattern: stars and atoms, and atoms in stars.

Cosmological clouds

Stars are not the only source of information concerning chemical complexification of matter through the ages. Clouds floating between the stars (interstellar) or between galaxies (intergalactic) selectively absorb certain notes of light and are thus susceptible to spectroscopic analysis. The variation in composition of cosmic clouds as a function of their redshift (which is a measure of their distance in space as well as in time) provides a prime source of information concerning the chemical evolution of the cosmos.

It is thus crucial to build up an inventory of metal contents at different epochs throughout the evolution of the Universe and everywhere across the Universe. This includes remote galaxies and also the great absorbing clouds on the line of sight to quasars, which serve as genuine cosmic beacons. One aspect of the current endeavour to trace back the large-scale history of the Universe is to determine the rate of growth of the metallicity Z as a function of time, or rather the star formation (astration) rate, and its variation as time goes by. For it is no secret: stars are the mother of metals.

The motivation is the same as that which launched spectroscopy in such a spectacular way over the past thirty years, that is, to test the theory of nucleosynthesis and the chemical evolution of galaxies. Cosmologists are delighted to point out that today we may assess chemical abundances in the remote Universe by pure observation, whereas even ten years ago, such a feat remained out of reach.

We owe the present rich harvest of observations to a whole new generation of astronomical instruments deployed around the Earth or on its surface. Giant telescopes equipped with 8-m or 10-m mirrors (VLT and Keck) are gradually replacing 4-m telescopes, such as the outstanding CFHT. The future is likely to be no less brilliant with Gemini and the New Generation Space Telescope in the offing.

For its part, the VLT is Europe's window on to the Universe. Data from other galaxies and extragalactic clouds are raining in like a spring shower and the cosmic archives are hastily being reorganised. Astronomers who for many years had grown accustomed to measuring abundances in stars and nearby galaxies are sometimes disconcerted by the diversity of the measurements and the occasionally unexpected nature of the results. But patience! We are just beginning the chemical inventory of the Universe.

Abundance measurements at high redshifts complement compositional studies of stars in our own Galaxy. These new measurements are like samples taken from different cosmic epochs. As we might have expected, they show an overall trend towards metal enrichment. However, the data are as yet fragmentary and fitful (see Chapter 8).

Exegesis of the abundance table

Let us return for a moment to more modest proportions. Indeed, let us return to the fold, to the friendly confines of the Sun, or more intimately still, to our fathering protosolar cloud.

Today we possess an abundance table of elements and isotopes characterising our local galactic environment, viz. the Solar System. We may draw several lessons from this document, a kind of Rosetta stone of nuclear astrophysics. A knowledge of the basic ingredients of the Universe serves to test and constrain one of the most fundamental theories of astrophysics, namely, the theory whereby the elements are nucleosynthesised by stars, supernovas, the Big Bang and galactic cosmic rays.

An economic and biological standpoint

Our task will be to give a general astrophysical meaning to these columns of numbers, to read the cosmic message they may contain, and then to proceed to a critical interpretation of this primordial diagram. But first let us approach it from a more human standpoint, indeed from a metallurgical point of view.

The jeweller knows full well why gold, with atomic number 79, is so precious. It is costly because it is rare. In fact, the Universe seems to have had great trouble in producing it. On further examination, we find that it is only synthesised in the innermost layers of supernovas, where iron is transmuted under an extremely intense flux of neutrons.

In contrast, iron peaks on the abundance table. The effect is too pronounced to be a chance effect. Indeed, iron stands apart in having the highest binding energy of all nuclei in creation, as any nuclear physicist will tell you. In the nuclear aristocracy, the highest place is held by this rather modest-looking metal and its kin, dreadnought nuclei able to resist the highest temperatures. The king of nuclear creation is therefore iron and the role of the stars is to pave the way to its long reign. Shall we proclaim it in the streets?

The stars produce these favoured nuclei in their fiery depths. But do they not then seize them like hostages for subsequent destruction? Is the king not held prisoner? Has there not been sufficient time to achieve nuclear perfection? These questions demand an answer.

Our chemical environment imposes constraints which can be summed up by Fred Hoyle's three rules:

1. Any material used in large quantities by the human species must be made up of elements that are extremely abundant in the Universe.
2. Less abundant elements may have economic importance, provided that their use is restricted to technologically advanced items in which small quantities of materials with very specific properties play the key role.
3. Elements with extremely low abundance can also be important economically if they are sources of nuclear energy, because the energy yield from nuclear processes is a million times greater than from chemical processes.

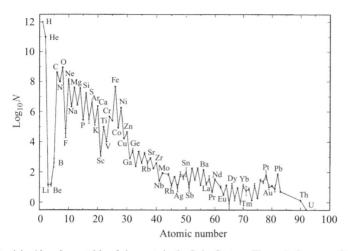

Fig. 4.1. Abundance table of elements in the Solar System. The main features of the abundance distribution are as follows: (1) the hydrogen ($Z = 1$) peak, shouldered by helium ($Z = 2$); the precipitous gorge separating helium and carbon ($Z = 6$); (3) the continuous decrease from the carbon region, i.e. from oxygen ($Z = 6$–8), to calcium ($Z = 20$); (4) the scandium ($Z = 21$) valley followed by the iron ($Z = 26$) peak; (5) the sawtooth landscape sloping gently down towards the small platinum ($Z = 78$) and lead ($Z = 82$) rises; (6) the flat lands of thorium ($Z = 90$) and uranium ($Z = 92$).

From the standpoint of the biologist, the atoms of life, carbon, nitrogen and oxygen, are strongly represented in the table. As we said before, this is true throughout the Galaxy. The ingredients of life are amongst the most equally apportioned in the world (the Universe).

There remains one burning question: what nature has done here on Earth, has it not done the same for others elsewhere in the Galaxy? The atoms of life are not limited to the Solar System. They are to be found scattered across the whole galactic disk. What is more, most of the stars in the disk have much the same composition as the Sun. The same proportions of carbon, nitrogen, oxygen, phosphorus, iron and so on prevail throughout the plane of the Galaxy. In fact it seems that the overall metallicity grows as we approach the galactic centre. The question is there, but we shall leave the answer to others.[2]

A cosmic and mathematical standpoint

Moving away from the terrestrial and anthropological context, it is immediately clear from the graph in Fig. 4.1 that hydrogen ($Z = 1$) and helium ($Z = 2$) are

[2] It seems to me that Jean Schneider and co-workers are among those who have most fruitfully tackled this question, at least in France.

by far the commonest elements, followed by oxygen ($Z = 8$), carbon ($Z = 6$) and nitrogen ($Z = 7$). For every 1000 atoms of hydrogen, there are 100 atoms of helium and about one atom of oxygen. Hydrogen comes from the Big Bang and oxygen from the stars, whilst helium is of mixed origins, although a large component is primordial.

Leading astrophysicists Fred Hoyle and William Fowler devoted their lives to solving the double enigma of the origin of the elements and the nature of the stars and supernovas which assemble atomic nuclei. Fowler was fascinated by the very small, by nuclear physics and its role in explaining energy production and the manufacture of new isotopes in stars. Hoyle was certainly interested in this problem, but his aim was to fit nuclear astrophysics into the larger framework of cosmology. In the mid-1950s, the two scientists engaged upon a fruitful collaboration, consecrating the idea that the Universe as we know it really is the seat and the result of nuclear processes. This idea would culminate in the publication of what is known in the trade as B^2FH, the founding article of nuclear astrophysics, coauthored by Margaret Burbidge, Geoffrey Burbidge, William Fowler and Fred Hoyle. This article marked the beginning of a new age. Vague speculations gave way to genuine arguments concerning known and well-observed stars in the sky.

The stellar brothers

One day in 1978, I asked 'Willy' (William Fowler) what roles he and Fred Hoyle had played in the marvellous theory of stellar nucleosynthesis which had set the stage for three generations of scientists. Finding the question difficult, he did not reply immediately, but several years later brought to my attention the scrupulously observed chronological description of the facts reproduced below.

The question was indeed a serious one which came to a head before the jury of the 1983 Nobel prize. Their conclusion was radical. The wise men separated the two stellar brothers on the grounds of scientific propriety. Fred professed ideas that were generally considered to be heretical, regarding the Big Bang and the origins of life. He was therefore deprived of the highest reward. However, he was later honoured with the prestigious Crafoord prize, which he shared with Salpeter. Fred Hoyle was one of the great visonaries of the twentieth century. In many respects, he was ahead of his time, but some of his work was not uncontroversial. He died recently. A symposium on 'Frontiers of Astronomy' was held in Cambridge (UK) in April 2002 to celebrate one of the founders of modern astrophysics.

In the following section, I summarise the contents of the letter in which William Fowler replied to my query. He began by recommending articles written by Fred Hoyle and Margaret and Geoff Burbidge in *Essays in Nuclear Astrophysics*, edited by Barnes, Clayton, and Schramm and published by Cambridge University Press. He also mentions brief references to the subject in the biographical memoirs of the Nobel prize (1983).

The letter may be considered as a genealogical document. Jean Audouze, Elisabeth Vangioni-Flam and myself belong to this line of descent. Hubert Reeves is the link between the pioneers and the current generation which is already preparing the succession.[3]

Letter from William Fowler to M. Cassé

What then was the sequence of events that led to the publication of B^2HF? They may be summarised chronologically as follows.

1946–9 Hoyle established the main lines of the synthesis of the elements up to iron in the stars. Others had put forward the general idea in a somewhat nebulous way, but Gamow's idea of complete nucleosynthesis in the Big Bang was generally accepted.

1949 Work carried out in the Kellogg Laboratory confirmed the existence of a gap at atomic mass number $A = 8$. Earlier, in the same laboratory, in 1939, the absence of a stable nuclei at $A = 5$ had been confirmed. Gamow was convinced that all the research relevant to atomic mass 5 was mistaken.

1951 During a visit to Kellogg, Ed Salpeter suggested a way that masses $A = 5$ and 8 could be bypassed in red giants via the reaction $2\alpha \Leftrightarrow {}^8Be(\alpha, \gamma){}^{12}C$.

1953 On the basis of studies of the HR diagram in collaboration with Martin Schwarzschild, Hoyle predicted the existence of an excited state of ${}^{12}C$ which would serve as a resonance for the reaction $2\alpha \Leftrightarrow {}^8Be(\alpha, \gamma){}^{12}C^*(\gamma){}^{12}C$. Whaling and his colleagues at Kellogg found this state very close to the energy predicted by Hoyle. At this point, Fowler began to believe in Hoyle's ideas about stellar nucleosynthesis.

1954–5 Fowler took a sabbatical year at Cambridge (UK) to work with Hoyle (financed by Fulbright and Guggenheim research grants). There he met the Burbidges and with them wrote two papers on what was later to be called the s process in B^2HF. Hoyle was busy with other problems. At the end of 1955, the Burbidges returned to Pasadena with Fowler, and Hoyle joined them at the beginning of 1956.

1956 In April, the results on the radioactivity found in the debris of the first atomic bomb (Eniwetok, November 1952) were finally declassified. Seaborg and his

---

[3] Jean Audouze and myself share the double honour of having prepared our doctorates under the supervision of Hubert Reeves and then beginning our postdoctoral careers with William Fowler at the California Institute of Technology.

colleagues found isotopes of californium ($Z = 98$) in the debris and this showed that the rapid capture of neutrons was capable of constructing heavy elements despite their fast α and β decay. The Burbidges, Fowler and Hoyle realised that the r process could develop from nuclei at the iron peak and get past lead ($Z = 82$) and bismuth ($Z = 83$), whereas the s process was blocked by these nuclei, to reach thorium ($Z = 90$), uranium ($Z = 92$) and their short-lived progenitors. Suess and Urey published their new determination of the elemental and isotopic abundances in the Solar System, showing indisputable evidence (double peaks) for the intervention of r and s processes. An article referred to as HFB[2] was published in the 5 October issue of *Science*, putting forward the two neutron-capture processes. Few people paid any attention.

1957 An extensive elaboration of HFB[2] was published in the October issue of *Reviews of Modern Physics*. This was the now famous B[2]FH. Al Cameron had already published his article in the June issue of the *Publication of the Astronomical Society of the Pacific*, but they were unaware of his work when they had submitted their manuscript to *Reviews of Modern Physics* at about that date. Cameron's article referred to HFB[2] but Fowler points out that it was nevertheless a remarkable feat for one person to arrive at the same conclusions as four who had worked very hard and in concert. B[2]FH attracted much attention and convinced people that nucleosynthesis in stars had been understood quantitatively.

Further developments had to wait for the appearance of the 1987 supernova and its detailed observation across the whole wavelength spectrum. Only then did nucleosynthesis become an observational science in its own right.

In the distribution of elements and isotopes that make up the Solar System, we are seeing the legacy of the stars. In the general economy of the Universe, stars play the role of conscientious craftsmen. Viewed purely in terms of productivity, they can be considered as heavy-element factories. In this respect, supernovas are of paramount importance.

Astrophysicists have built up models with equations rather as birds build their nests with twigs. And from them spring suns and supernovas. It has required much patience and years of reflection.

In order to model a stellar explosion, much work is needed. At the end of the day, we may read off the balance sheet, as it were. Depending on its mass and its metallicity at birth, we may predict the final account for each star. We admire the generosity and diversity of these stellar production units. When it explodes, a typical massive star delivers up around 0.1 solar masses of iron, and 2 solar masses of oxygen (see Appendix 4).

The French poet Prévert tells us that numbers are birds and algebra is in the tree tops. For our part, we may say that the stars do arithmetic. The star is the ultimate furnace in the art of nuclear alchemy. It is a place where the simple is made complex by adding together nucleons at high temperatures. Helium is the result of a fourfold union between hydrogen nuclei. Carbon comes from a threefold union of helium nuclei, magnesium from a twofold union of carbon nuclei. The father of iron, nickel-56, ultimately arises from a manifold combination of helium nuclei. The stars add up nucleons to multiply the numbers of nuclear structures:

$$4 \text{ hydrogens} = 1 \text{ helium } (4 \times 1 = 4)$$
$$3 \text{ heliums} = 1 \text{ carbon } (3 \times 4 = 12)$$
$$2 \text{ carbons} = 1 \text{ magnesium } (2 \times 12 = 24)$$
$$14 \text{ heliums} = 1 \text{ nickel-56 } (14 \times 4 = 56).$$

Genesis would thus appear to be resolutely arithmetical.

Discursive analysis of the abundance table

After going through this exercise in pure spontaneity, let us carry out a harsher analysis of our fundamental data. Abundances in the Solar System reveal trends that directly reflect not the chemical or atomic properties of elements, but rather the characteristics of the nuclei of those elements. The key to understanding the abundance table thus lies in nuclear physics.[4]

Nuclear alchemy

Suddenly it seems that our flight has carried us too far and too quickly. Let us return for a moment to the warm summer of the ancient world. Speak softly now under the olive trees and the stars!

Democritus thought there was an eternal form of matter, composed of indestructible atoms of many and varied types. But Aristotle found it absurd to suggest that the elements might be composite. His preference was for the existence of just four fundamental substances – fire, water, earth and air – governed by four states. Aristotle taught that the dry and the cold combine to form earth, the cold and the damp to form water, the damp and the hot to form air, and the hot and the dry to form fire. This theory made it possible to move from one element to another via their common attribute. The material world of Aristotle allowed for the mutation, and indeed the permutation of the elements. On the basis of this erroneous theory, alchemists built up their own understanding of the nature and existence of matter. They hypothesised their own transmuting agent, the so-called philosophers' stone, which, if found, could transmute

---

[4] See, for example, the article by R. Lehoucq and M. Cassé in *Pour la Science*, August 2000.

all base metals into gold and cure all illnesses by providing the elixir of life.

Following the philosopher's teachings, alchemists would sublime and distill, brew and grind. By amalgamating and scorching, they hoped they would obtain gold from one of the four elements. But never was gold to glitter upon the doorsteps of their laboratories.

Today this research programme raises a sarcastic smile. However, the basic idea behind it, whereby all forms of matter have a common origin and can transmute from one form to another, lines up well with the contemporary notion of a unified theory of matter. In this, science owes something to alchemy. In their untiring quest for gold, the alchemists subjected every known substance to the test of fire and acid (*aqua regia*), thus paving the way to modern chemistry.

However, the secret of transmutation did not lie in chemistry and the peripheral electrons that determine the chemical properties of the atom. Instead, the solution to this mystery had to be sought in the nucleus of the atom and the strong and weak nuclear interactions which organise and structure it.

The physical and chemical properties of an atom are determined by the number and configuration of electrons in its electronic retinue. These are arranged in layers or shells, in a well-defined order. Some atoms have more shells than others, or indeed their shells are more complete and better organised. Chemical properties and molecule formation are determined by the outer shell. This is because only the outer electrons can mediate in chemical bonds, playing the role of a common currency. Atoms in the first column of Mendeleyev's periodic table have a single electron in their outermost shell, whilst those in the second column have two, and so on, until we reach the noble gases which have eight electrons in their outer layer (except for helium, which has two).

Two electrons can be housed in the first shell. As the helium atom already contains two, the hotel is full and helium will not take in a further electron belonging to the outer shell of another atom. It will not therefore stick on to a fellow atom to form a molecule, whence its chemical lethargy.

The next atom on the list is lithium. It has a third electron, which must be placed in the second shell. This second shell can put up a total of eight electrons. Lithium thus contains two electrons in its first shell, now saturated, and a single electron in its second shell, the outer shell. This gives it a likeness to hydrogen, with a single electron in its outer (first) shell, held out like a hand to other atoms.

Oxygen has eight electrons, two on the ground floor and six on the first. There remain two empty rooms on the first floor, which can house at most eight electrons. Oxygen is therefore ready to accept two further tenants from elsewhere. For example, it is ready to share the electrons of two hydrogen atoms. The result is the water molecule H_2O. It takes two hydrogen atoms by their outstretched hands and forms H–O–H. The sharing of electrons creates a chemical bond.

Chemistry is the art of combining atoms. Nuclear physics is the science of the transmutation of the elements. Bombarding atomic nuclei by other atomic nuclei can produce transmutation of both target and projectile. Alchemy is thus nuclear, not atomic.

Abundance and scarcity

The high abundance of helium-4 is so striking that we cannot help wondering why certain heavier nuclei, more stable and better constructed, have remained so scarce. Iron is a case in point. Why does the work of nuclear transmutation have this unfinished air about it?

The simplest elements originating in the Big Bang are hydrogen and helium, comprising 98% of the total mass. The predominance of low mass nuclei ($Z = 1$–4) compared with high mass nuclei such as nickel-56 (atomic mass number $A = 56$) is a consequence of two factors: the omnipresence of photons in the original soup and the instability of helium's ailing progeny (see Appendix 1), that is, nuclei of masses 5 to 8.

The simple is generally abundant, and the complex rare. Returning to our earlier question, the reason why gold is so rare is basically electrical. One obvious inhibiting factor when building up complex species is the electrical repulsion between nuclei. The probability that two nuclei join together decreases exponentially with the product of their electrical charges. For example, the fusion of two carbon nuclei involves a product 36 times greater than for two hydrogen nuclei, and this number occurs as the argument of an exponential. In other words, the probability is a factor of $1/\exp 36$ lower. Such strong inhibitions can be overcome by increasing the relative velocity of the nuclei, that is, by increasing the temperature, since the latter is merely a measure of thermal agitation. But the required high temperatures are not to be found at every turn.[5] The same can be said for the high neutron fluxes that build up extremely complex nuclei in very tiny steps.

What then of the clouds of nuclei billowing out from the stellar furnace? One notable feature is that nature seems to prefer even numbers to odd. Apart from light hydrogen ($A = 1$), which is indeed a very special case, nature clearly favours the even. Abundances thus feature a marked even–odd imbalance. For those nuclei with fewer than 20 protons, the most abundant isotopes contain the

---

[5] See, for example, the article by Ludwick Celnikier published in M. Cassé, *Nucléosynthèse et abondance dans l'univers* (Cépaduès, Paris, 1998).

same numbers of protons and neutrons. For nuclei with Z between 20 and 30, the most abundant isotopes are those with $N = Z + 2$ neutrons. For higher Z, the most abundant isotopes are richer in neutrons than protons.

Let us note the following key point: nuclei for which Z and N are even are more abundant than their immediate neighbours. Nuclei with even atomic mass number A are favoured to the detriment of those with odd A. We must move down to the fifteenth position on the list of nuclei in order to find the most abundant one with odd mass number after hydrogen. This is magnesium-25. Note also the sudden drop in abundances in the region of $A = 5$–11 and around $A = 45$.

The fundamental reason why an even number of protons and an even number of neutrons is favoured is that a solitary nucleon, frustrated by a lack of relationship with one of its own kind, can only be harmful to nuclear stability.

Only ^{14}N amongst the abundant nuclei is not even–even. This exception can be explained by the fact that nitrogen occurs among the ashes of the first step of thermonuclear fusion in stars, to wit, the fusion of hydrogen, the commonest one could imagine.

Apart from ^{56}Fe, the majority of the frequently occurring nuclei are even–even and also have the same number of protons and neutrons ($Z = N$). The most abundant are ^{16}O and ^{12}C, followed by ^{20}Ne, ^{24}Mg, ^{28}Si, ^{32}S, ^{36}Ar and ^{40}Ca.

Other configurations are also favoured, in particular, and doubly so, those that bring together a proton number and a neutron number equal to 2 (^{4}He), 8 (^{16}O), 20 (^{40}Ca) and 28 (^{56}Fe, which is actually produced via ^{56}Ni).

In fact, the stars show a special partiality towards ^{56}Ni. It is the species towards which all others converge at temperatures above a few billion degrees, that is, at the kind of explosive temperatures prevailing in supernovas.

Iron, immediately evident by its peak on the abundance curve, owes its existence to the fact that the most robust of all known nuclei is born in the form of radioactive nickel-56. The resplendent light of supernovas, gleaned from the transmutation of nickel into iron, proclaims its illustrious birth in the heavens, as we shall see in more detail shortly.

The precipitous chasm inhabited by lithium, beryllium and boron reflects the extreme fragility of the nuclei of these species. Note that fluorine, located just above the favoured number of eight protons, is as expected rather scarce.

Beyond iron lies a first population of so-called s-process nuclei, which includes among others barium and lead. This population has an abundance distribution with peaks around mass numbers 87, 138 and 208. These nuclei are produced by slow neutron capture, referred to as the s process. A second population, slightly shifted from the first, including gold, platinum and uranium, is imputed to the process of rapid neutron capture, referred to as the r process.

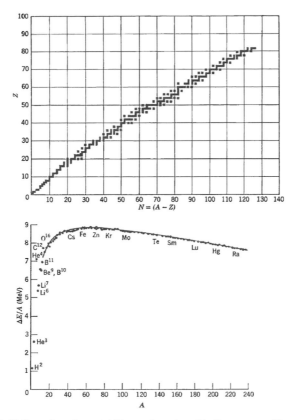

Fig. 4.2. Valley of nuclear stability and nuclear binding energy. *Top*: Beyond $Z = 20$, the distribution of stable isotopes curves downwards in the (N, Z) plane, showing that stable nuclei grow richer in neutrons as their atomic number increases. *Bottom*: The binding energy per nucleon, $\Delta E/A$, is a measure of how robust a nuclear species is in the face of attempts to break it up. This curve reaches a peak around iron.

Nuclear stability and interactions

Extensive studies of nuclear reactions and the systematic behaviour of nuclei played a central role in developing a theory of the origin of the elements. The idea was to reproduce the abundances observed in the Solar System, an aim which inevitably raises the following question.

Assuming that all combinations of neutrons and protons can exist, which atomic nuclei are stable enough to survive for as long as the Universe itself, that is, for around 10 billion years? Estimates of nuclear stability are available to answer this query. The 270 or so nuclei found in nature in some lasting form all lie along what is known as the valley of stability in the (N, Z) plane (Fig. 4.2).

It appears that the most stable light nuclei comprise equal numbers of protons and neutrons, i.e. $Z = N$, whilst heavier stable nuclei involve an excess of neutrons which act as a buffer against increasing electromagnetic repulsion between protons within the nucleus.

Nuclei with the highest binding energy are the most difficult to destroy and hence the most stable. Note that iron (^{56}Fe) stands at the peak of the nuclear stability curve. It therefore emerges as the heir apparent of nuclear creation, so to speak, although history has decided otherwise. In any case, the exceptional nuclear stability of iron is of great importance for the synthesis of heavy elements, serving as a stepping stone.

The stability curve is a powerful predictive tool with regard to the emission or absorption of energy in a given nuclear reaction. The figure can be used to estimate the tendency of a nuclear reaction to go one way or the other. For example, fusion reactions between light nuclei generally lead to more stable nuclei and thus release energy. Fusion reactions producing nuclei heavier than iron manifest the opposite tendency, thus requiring an input of energy.

As a consequence, by breaking up a heavy nucleus (fission) to produce lighter and more stable nuclei, energy is once again released. Detailed examination of the stability curve reveals other important features of nuclear structure which are strongly correlated with observed abundances.

For example, in the case of relatively light nuclei, structures composed of several helium nuclei, such as ^{24}Mg, ^{28}Si, ^{32}S and so on, are more stable than their neighbours. Moreover, nuclei with even mass numbers (an even total number of neutrons and protons) are more robust than those with odd mass numbers.

Nuclear ninepins

Once we have decided whether a nuclear reaction will release energy or absorb it, the next step in ranking it amongst all possible reactions is to assess the probability with which it can produce the sought result. Given that the temperature at the centre of the Sun is about 15 million K and the density 150 g cm^{-3}, we would like to know how long it will take our star to consume all its nuclear resources.

This probability depends on the reacting nuclei and prevailing physical conditions, such as temperature and density. It is easy to understand the relevance of the density: the more particles there are per cubic centimetre, the more frequent are the collisions. The role of the temperature is also fundamental. Nuclear reactions require high temperatures, in fact, all the higher as the reacting nuclei carry higher electrical charges, as mentioned above.

On the most basic level, since nuclear reactions result from collisions, the probability of formation of a given nuclear species depends on the target, the projectile and their relative velocity. Amongst all possible combinations involving as projectile either protons, neutrons or photons and as target any nucleus from hydrogen to uranium, what are the most productive reactions? In fact they are simply the most probable ones.[6]

The number of nuclei of a given species produced by a reaction of type

$$1 + 2 \longrightarrow 3$$

in 1 second and 1 cm^3 can be assessed as it would be in a game of ninepins. Indeed, the number of successful shots is just the number of balls thrown multiplied by the number of pins multiplied by the sum of the cross-sections of pin and ball.

The only major difference is that the diameter of the ball in nuclear reactions must be treated as velocity dependent. A parallel can be drawn with the wavelength $\lambda = h/mv$ of the projectile particle as given by quantum theory. The probability of the reaction thus depends on the relative velocity of the reactants.

Reaction probabilities have been extensively measured in many laboratories all round the world, and in particular at the California Institute of Technology under the guidance of William Fowler. But whenever experimental data is lacking, nuclear reaction rates can be estimated theoretically. In this way, reaction rates have been fully tabulated as a function of temperature, ready to be integrated into numerical codes for stars or the Big Bang, and made available to the whole astrophysical community.

It should be emphasised that a reliable and up-to-date database of nuclear reactions is of capital importance for theoretical attempts to understand the origin and evolution of the elements. Calculations of the nucleosynthesis occurring in the Big Bang, the Sun, the stars and supernovas depend critically upon it.[7]

On the basis of these estimates, we can identify the flow of nuclear reactions and plot the rivers they follow on the (N, Z) map. By coupling this network of nuclear reactions with models of stars or the Big Bang, which predict temperature and density variations in space and time, we may hope to identify the nature of the elements and isotopes produced, as well as their relative proportions.

---

[6] The key parameter here is the probability that a projectile A with speed v colliding with a target B should produce a nucleus C, among other things. This is given in the form of a parameter known as the reaction cross-section.

[7] I would like to pay tribute to Marcel Arnould at the Université Libre in Brussels, a master craftsman in this art, to the Orsay experimental group in Paris (CSNSM/IN2P3) who work on the determination of reaction probabilities of astrophysical import, and to Alain Coc, in particular, and to all the physicists at the GANIL installation (Grand Accélérateur National d'Ions Lourds) in France.

The abundance curve of chemical elements was used by Burbidge, Burbidge, Fowler and Hoyle in 1957 and by Cameron in the same year to establish the basic process of nucleosynthesis operating in stars. The breakdown of mechanisms responsible for the existence of the various types of atom in the observed proportions is as follows:

- hydrogen fusion in stars: slow conversion of hydrogen into helium at temperatures above 10 million K, over time-scales of the order of 10 billion years;
- helium fusion: conversion of helium into carbon, oxygen and so on, at temperatures above 100 million K, over time-scales of the order of 10 million years;
- carbon fusion: production of neon, magnesium and aluminium at temperatures above 600 million K;
- oxygen fusion: production of nuclei between oxygen and silicon at temperatures above 1 billion K, over time-scales of 100 000 years, unless this nucleosynthesis is explosive, in which case it lasts only a few seconds;
- silicon fusion: production of nuclei between silicon and zinc, at temperatures above 3 or 4 billion K, on a time-scale of a few hours in the non-explosive case and less than 1 second in explosive conditions (Fig. 4.3);
- s process: slow neutron capture, responsible for the synthesis of various nuclei with atomic numbers above iron and requiring temperatures above 100 million K and neutron fluxes sustained over a thousand years to 100 million years.
- r process: rapid neutron capture, producing nuclei with atomic numbers above that of iron and requiring temperatures above 10 billion K over very short time-scales;
- p process: production of rare species, rich in protons, at temperatures of 2 to 3 billion K over 10 to 100 seconds.

In addition to all these fusion and neutron capture processes, there is a further type of nuclear reaction, called spallation. Rather than fusing together, nuclei are smashed up or chipped to produce smaller species. This process is thought to be the origin of most of the lithium, beryllium and boron in the Universe.

We shall return to these mechanisms for creating ever more elaborate nuclei. But first, let us glance back at the abundance table, for it would be an oversight not to mention the rich nuggets of information that lie hidden in meteorite studies.

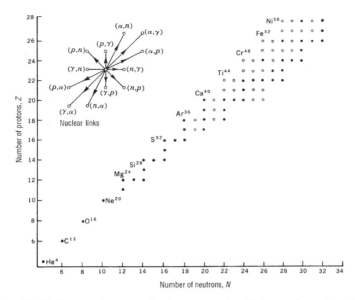

Fig. 4.3. Nuclear reaction network. As an example, this figure shows the links established using nuclear reactions as a go-between during silicon fusion. Black dots indicate stable isotopes. The notation (x, y) is equivalent to $x + A \rightarrow B + y$, where x is the initial particle and y the final particle. (After Clayton 1983.)

Supernova dust

Held in meteorite magmas, we find grains or granules called chondrules, whose compositions deviate from the average. Indeed, they exhibit isotopic anomalies with respect to the canonical abundance table. These tiny inclusions represent genuine mineral relics from the stellar age. Although the greater part of the matter in the Solar System displays a high degree of isotopic homogeneity, which is just what allows us to build up the abundance table discussed earlier, a small fraction of this matter (one ten-thousandth, perhaps) is characterised by a quite different range of isotopic compositions from the average.

The Allende, Murchison, Murray and Orgueil meteorites are particularly highly prized for research into stellar grains, since several kilograms of this material have been identified in each of them. This is sufficient to be able to take samples of the order of 1 g without damaging the source. Such samples can then be subjected to compositional analysis. But how can we extract these stellar jewels, measuring at most 1 μm in diameter, from the matrix in which they are embedded? The best way of finding a needle in a haystack is to burn the hay. Cosmochemists employ basically the same method when they use chemical processes to isolate star dust trapped in meteoritic stone. They may then analyse it at their leisure via the tried and tested methods of geochemists. Ion probes and mass spectrometers thus become the instruments of astronomy.[8]

Isotopic ratios such as the ratio of carbon-12 to carbon-13, corrected for physicochemical enrichment or impoverishment, are imputed to purely nuclear, and hence stellar, processes.

Today, through astronomical observation and stellar models, we know that most galactic dust (about 1% of the mass of interstellar clouds) is produced by red giants at the end of their lives, or rather, in the particular evolutionary phase when they embark upon the asymptotic giant branch of the HR diagram. When a star grows old, carbon builds up in its envelope. Convection currents carry the carbon up from the depths where it is synthesised by triple capture of helium nuclei in the star's hot core. Carbon wins out over oxygen and a carbon-bearing dust forms (silicon carbide). The red giant is now spattered with grains whose isotopic composition carries the fingerprint of events occurring deep down within the star.

The different types of grain can be related to specific classes of stellar objects. The very hot and bright, even lavish Wolf–Rayet stars are considered to be one of the most favourable sites for grain formation, for their strong stellar winds are particularly rich in carbon. Matter thrown out by supernovas and cooling very quickly due to its expansion is also an excellent scenario for grain formation. Elements with any affinity for the solid state are likely to be abundantly transformed.

Whilst the highly canonical abundance table reflects the final result of a great many nucleosynthesis phenomena, involving contributions from quite different sources (but mainly supernovas, red giants and planetary nebulas), stretching across some 10 billion years that preceded the birth of the Sun, it is thought that the isotopic anomalies result from pollution of the protosolar nebula by a few isolated sources. They thus inform us about a small number of special stars which came along to put a few finishing touches to abundances in the Solar System. Stellar grains inserted into celestial stones thus provide further information about nucleosynthesis and the nature of their source stars. Astronomy is also written in stone.

It was indeed a pleasant surprise to discover that certain grains carrying the unmistakable signature of supernovas had survived the turbulent formation of the Solar System. In addition it proved possible to extract them from meteorites without modifying them, so that they could be studied at leisure in terrestrial laboratories. Dissolving the stony component in acid and carrying out a series

---

[8] The astrophysical community is very grateful to François Robert and Marc Chaussidon for their contributions to isotopic analysis of meteoritic matter.

of oxidations and separations, the isotopically anomalous components were isolated.

Supernova grains discovered in meteorites are characterised by anomalous proportions of oxygen-16, magnesium-26, silicon-28 and calcium-44. The excesses of magnesium-26 and calcium-44 are produced by radioactive decay of aluminium-26 and titanium-44, respectively, within grains formed in the burning envelope of the supernova. Today, we may hold in our hands tiny solid particles that came into being before the Sun itself. Presolar grains are non-biological fossils of extraordinary cultural and scientific importance. They were born in the debris, the winds and the fragments of stars. They have spent untold ages wandering through space before being incorporated into a protosolar cloud.

Unlike stellar spectroscopy, the analysis of meteoritic grains and inclusions can provide an extremely precise isotopic breakdown. The weak point of this technique, however, is that the exact characteristics of the stars from which the grains formed can only be inferred. When we detect light, we can deduce its celestial source by extending back its line of incidence and we can determine the composition of the source from the spectral lines it contains. But we do not know where the meteorite grains came from, and only their composition can tell us anything of their origins.

Each known type of grain is made from a particularly refractory form of material. Themselves born in extreme heat conditions, these grains survived the formation of the Solar System without the least difficulty. They have been able to carry down the isotopic composition of their source quite intact, throughout the whole prehistory of the Sun. But their message has not yet been perfectly decoded. The story of this star dust will therefore be continued, especially as it is radioactive and can be identified by its gamma emissions.

Astronomy of radioactivity

The best illustration of radioactive astronomy is titanium-44. We shall take it as the archetype of a good radioactive isotope. It is relatively abundant and has a reasonable lifetime of around 100 years, neither too long, nor too short. Only aluminium-26 can rival it in this respect and nuclear gamma astronomy has already reaped some of the rewards (see Fig. 4.4).

Calcium-44 bears the same relation to titanium-44 as a grandson to his grandfather. The radioactive affiliation is

$$^{44}\text{Ti} \longrightarrow {}^{44}\text{Sc} \longrightarrow {}^{44}\text{Ca} .$$

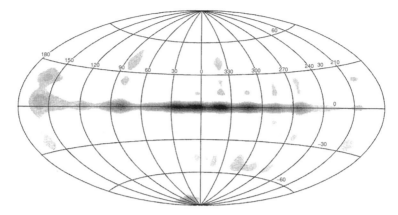

Fig. 4.4. All-sky map in the light of the 1.809 MeV gamma-ray line from radioactive aluminium-26. The galactic distribution of aluminium-26, based on data from the COMPTEL (Compton Telescope) experiment aboard the GRO (Gamma-Ray Observatory), suggests that this isotope is dispersed across the Galaxy by the most massive stars, Wolf–Rayet stars and supernovas. ^{26}Al is formed by the reaction ^{26}Mg + p → ^{26}Al + γ. This radioactive isotope has a lifetime of about million years and is ejected into space before it begins to decay.

Titanium-44 transmutes to scandium-44 by emitting two gamma rays, at 68 and 78 keV. The new nucleus then transmutes to calcium-44 but not before emitting a gamma ray at 1.157 MeV.

The search for titanium-44 was undertaken by the gamma spectrometer aboard the Gamma-Ray Observatory (GRO). The 1.15 MeV line was detected in the direction of Cassiopeia A and Vela, two recent supernova remnants. Mapping the Galaxy in the 1.15 MeV line will undoubtedly be one of the main objectives of the European satellite INTEGRAL, a unique space-borne experiment in which France is deeply involved.[9]

Titanium-44 is thus an isotope of considerable astrophysical importance. Detection of its characteristic gamma line has aroused great enthusiasm amongst the nuclear astrophysics community because this isotope supplies precious clues as to the explosion mechanisms operating inside massive stars. Indeed, it allows us to determine the exact boundary between the part of the star which is imploding, to end up as a neutron star, and the part which is ejected and flies out into space, loaded with atomic nuclei manufactured by the star. It also provides

[9] I would like to thank all the artisans of this huge project for their skill and self-effacing commitment, in particular, Jacques Paul, Bertrand Cordier, François Lebrun, Philippe Durouchoux and Jacky Grétolle at the Saclay Research Centre, and Gilbert Védrenne, Jurgen Knödlseder, Peter Von Ballmos, Pierre Mandrou, Jean Pierre Roques and the whole team in Toulouse.

information about the maximal temperature and density attained during passage of the shock wave concomitant with the explosion.

The second interesting feature of this isotope is that minuscule grains of silicon carbide extracted from meteorites have been found to be very rich in calcium-44, as mentioned earlier. They have been identified with presolar grains that condensed in the ejecta of supernovas during their first few years of expansion. Could it be that supernovas have been throwing sand in our eyes? Data gathered by the ISO (Infrared Space Observatory), yet another experiment with strong participation by the French CEA, clearly demonstrates that new dust condensed inside the Cas A remnant very soon after explosion of the supernova that caused it.[10]

[10] I would like to pay tribute to Catherine Cesarsky, Laurent Vigroux, David Elbaz and Pierre-Olivier Lagage for their infrared enthusiasm.

5

Nuclear suns

Glossary

alpha particle helium nucleus
astroparticle physics new term invented to denote the common ground between astrophysics and particle physics
Cerenkov radiation emission of blue light by particles moving faster than light in a medium other than the vacuum (note that in a medium with refractive index n, light moves with speed $v = c/n$)
helioseismology study of the solar interior by analysing vibrations at its surface
$M_\odot$ solar mass
main sequence curve on the HR diagram where stars are located when converting hydrogen to helium in their core
nuclear statistical equilibrium phase of high-temperature nucleosynthesis in which the abundance distribution of nuclei depends only on their robustness
photodisintegration process whereby nuclei are partially broken apart by photons
plasma ionised gas
shock wave effect associated with supersonic motion

The Sun as reference

The motions of the Sun and Moon form the basis of our calendars. The measurement of mechanical time is largely based on the periodic reoccurrence of certain phenomena: the rhythm of day and night, the seasons, or the cyclic reappearance of the planets and stars in the sky. The flow of change is attested by the apparently irreversible global evolution of the Cosmos. Cosmic time, eternity's yardstick, is the measure of universal change, of the evolution of matter, and this evolution is essentially one of nuclear complexification, driven by stellar forces.

The material evolution we are speaking of here is at work in all galaxies. Every part of the Universe is evolving, and the driving force is the stars. Everywhere on Earth, there are men, women and children; everywhere in the sky, there are stars. The star seems to be the best-adapted form of the visible Universe.

The path that leads from the multitude of anonymous and abstract elementary particles generated in the original explosion to the grass in the meadows, to the rain and the wind, to the infinite variety of shapes and states, to the profusion of feelings, must necessarily pass through the stars. Stars are an essential link between the primordial raw material that came out of the Big Bang and complex material with the ability to think. Nuclear astrophysics is the bridge between elementary particle physics and life.

Star, driving force behind the chemical evolution of the galaxies, mother of atoms and of all life, gentle or explosive, let us seek to become better acquainted. For it is one thing to observe and record the state of atomic matter in the Universe, and quite another to explain it. It is to this Herculean task that nuclear astrophysics dedicates it best troops. And the starting-point for each sally is the Sun, our personal reference.

One of the aims of nuclear astrophysics is to understand how nuclear processes generate energy in the Sun and stars, thus supplying and perpetuating their brightness. Another is to know how, in doing so, they succeed in synthesising complex elements from simpler ones, from primordial hydrogen and helium inherited from the Big Bang.

For astronomers, the Sun is a star, indeed, the closest star in the sky. We see only its luminous outer layer, the photosphere or sphere of light. It appears to us as a disk but is of course spherical in reality, with radius 700 000 km, calculated from its angular diameter and its distance. Its surface temperature is 5760 K, assessed via its colour. It has a total mass of 2×10^{30} kg, according to the calculations of celestial mechanics. It has an intrinsic luminosity of 4×10^{26} W if we go by the light flux measured above the Earth's atmosphere and take into account the distance from Earth to Sun. Then, apart from the electromagnetic activity occurring on the surface in the form of solar flares, sunspots, granulation and so on, which interest only solar dermatologists, the chemical composition of its atmosphere can be determined by spectroscopy.

Sensitive study of the Sun's light shows that the Sun is vibrating. In fact, it vibrates like a bass drum and its light emissions vibrate in symphony. The quivering of the Sun's surface teaches us something about what is inside it. Rather like the doctor who taps a patient's thorax with curved forefinger, we auscultate the Sun as we might test a coconut or a melon. Depending on how it reverberates, we can tell whether our fruit is full or empty. The Sun too clatters and rattles, although in this case no one is there to strike it. The blows are brought to bear from within. Indeed, solar oscillations are maintained by internal hammering due to the incessant motions of granules. We infer the speed of sound at various depths and proceed to compare it with the predictions of our models.

Is the Sun then eternal? And if not, what stage has it reached in its evolution? In fact, the Sun is in the simplest and longest-lasting phase of a star's evolution, the quasi-static phase of core hydrogen burning, referred to as the main sequence.

Are the stars liquid? Given that the Sun is spherical with radius 700 000 km, we may find its volume from the formula $V = 4\pi R^3/3$, then calculate the mass divided by the volume to deduce its average density. To our surprise we discover that 1 cm^3 of the Sun weighs 1 g. But this is exactly the density of water. Could the Sun be aqueous?

We sometimes hear it said that stars are incandescent balls of gas. It may indeed be surprising to find that a material with the density of water, such as the substance of the Sun and many stars, should be described as a gas.

The gaseous state of stellar matter is nevertheless a reality, and it is easy to see why. The atoms in it are dissociated by heat, and the entities thereby released, that is, electrons and atomic nuclei, are sufficiently far apart to be treated as free. The distance between the constitutive particles is much greater than their own dimensions. Such is the definition of a perfect gas. The company of these particles then displays an admirable flexibility and power of adaptation that ensure it a long life.

But what exactly causes this longevity? In fact, the day-star owes its long life and stability to the flexibility of its gaseous core. For although it has the same average density as water, it behaves as a gas because of the extreme heat levels maintained within it. Now gases have a rather special property, namely that they cool when they expand, and heat up when compressed. The perfect gas equation applies to the stellar fluid: the pressure multiplied by the volume is proportional to the temperature. This relation between temperature and volume is the key to solar stability. Hence, if a nuclear reaction occurs at the centre of the Sun, its core dilates. The temperature then drops correspondingly and the reaction slows down. If on the other hand a reaction weakens, less heat is produced and the core contracts. This causes the temperature to rise, encouraging the reaction to greater effort. Such self-regulation, related to the properties of the gaseous state, is what guarantees the Sun's longevity and stability.

The plasma phase

Under normal conditions, gases are electrical insulators, i.e. electricity cannot pass through them. However, by heating them to high enough temperatures, physicists discovered that they actually become very good conductors. They are in fact transformed into plasmas within which electrons are relatively free to circulate.

In an atomic gas, electrons are held captive by atomic nuclei. In a plasma, they are free to move around. This freedom of the now-autonomous electrons makes the plasma a good electrical conductor. The plasma contains a mixture of free electrons and positive ions.

Solid bodies resist compression just as they do expansion. The forces giving this coherence to solids are of electrical origins. They may be broken by heat, which communicates a disordered energy to the constitutive molecules of the solid. If the solid acquires enough energy in the form of heat, the molecules are freed, then broken into their component atoms. Any material can be brought into a gaseous state by putting in enough energy. What happens when even more energy is injected into our already vaporised system? Eventually, the internal structure of the atoms is shattered and the electron shells are gradually peeled off. When an atom loses an electron, it becomes once positively charged. If it loses two, it is twice positively charged, and so on, until the nucleus has been completely ionised, that is, all its electrons have been stripped from it.

Therefore, by adding enough energy to any material, we can eventually dissociate it into atomic nuclei and electrons, hence turning it into a plasma. Our own Sun, unlike the Earth and the other planets, is in this plasma state, as are all stars. Although plasmas on Earth are artificial, the greater part of the visible Universe is made up of plasmas.

The fact that the Sun is in the plasma state means that it has the flexibility of a gas. This flexibility in turn ensures its longevity. Indeed, the size of the particles making it up, i.e. separate nuclei and electrons, is much less than the mean distance separating them, and this allows us to identify it with a gas. (Non-baryonic dark matter, with no electrical properties, cannot be ionised. There is no such thing as a dark matter plasma.)

The most surprising thing about the Sun is not that there are still some doubts concerning the processes going on within it, but rather that we should have acquired such an accurate description of its inner workings. We are now ready to answer our children's questions.

What makes the Sun shine? The chemist's atoms would be unable to maintain the Sun's flame for very long. A phosphorus Sun on the end of a matchstick would only shine, at the Sun's present rate of energy release, for a few tens of thousands of years. But we know that the Sun, like the Earth, has been in existence for some 4.6 billion years. What is its secret? In fact, it lies in the nucleus of the atom. This is the grain the Sun grinds, or rather, agglutinates. The secret is numerical.

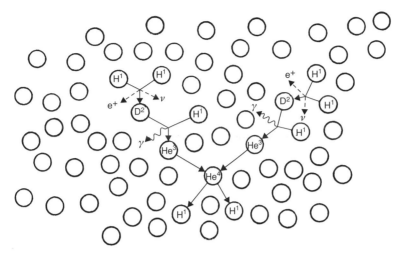

Fig. 5.1. Hydrogen fusion via the proton–proton chain.

Number one is hydrogen, with one proton, whilst four is helium, with two protons and two neutrons. Four protons fuse together under the influence of heat to give a helium nucleus, and this happens billions of times over. The mass of the helium nucleus is slightly less than the sum of the masses of the four protons. What has happened to the difference? In fact, it has been radiated away. This is why the Sun shines!

Hydrogen fusion

When the core of a star similar to the Sun reaches a temperature of 10–20 million K and a density approaching 100 g cm^{-3}, the protons in this hot, dense medium acquire enough kinetic energy for nuclear reactions to commence. Once triggered, the process of hydrogen fusion can proceed over very long periods (Fig. 5.1). The stars then remain at a fixed position on the main sequence curve of the HR diagram. Their exact location depends on their mass, the Sun being at an average position. Ninety percent of the stars in the Universe are at this stage in their lives, burning hydrogen and transmuting it into helium by the following reaction chain:

$$p + p \longrightarrow D + e^+ + \nu \quad \text{(twice)},$$
$$D + p \longrightarrow {}^3\text{He} + \gamma,$$
$${}^3\text{He} + {}^3\text{He} \longrightarrow {}^4\text{He} + 2p,$$

leading to a total reaction

$$4p \longrightarrow {}^4\text{He} + 2e^+ + 2\nu + 26.7 \text{ MeV}.$$

The synthesis of helium follows a somewhat indirect path. The longest part is the first, because it involves the transformation of a proton into a neutron, and such transmutations proceed via the weak interaction (slow at these temperatures). The leisurely pace of these reactions confers long life upon main-sequence stars.

The fact that neutrinos are emitted during the transformation provides an opportunity for direct observation of the reactions taking place at the heart of the Sun. Note that antimatter is produced in this strange reaction, in the form of the positron or antielectron e^+. The positrons generated immediately annihilate with electrons in the surrounding medium with subsequent emission of gamma rays.

Figuratively speaking, just as hydrogen is said to burn, we may say that helium is the ash left over from the reaction. The imponderable separates from the ponderous, photons and neutrinos take flight, leaving the heavy helium ash to accumulate gradually. The ample energy disgorged (26.7 MeV) makes this reaction chain one of the most generous known. For example, thermonuclear fusion of 1 g of hydrogen releases some 20 million times more energy than the chemical combustion of 1 g of coal.

The energy released by hydrogen fusion serves to stabilise the star against its natural tendency to collapse under its own weight. Nature has thereby set up one of its most beautiful equilibria. The nuclear fuel is abundant and rich, allowing it to shine for a long time with a sustained brightness. No upset can occur in this fine balance until 10% of its mass has been consumed as nuclear fuel. Its internal structure remains the same and the star is as though nailed to the spot on the HR diagram.

In stars with masses greater than 1.2 times the solar mass, hydrogen fusion proceeds via another channel, the so-called CNO cycle. This process tags together proton captures and β decays in the following chain of reactions:

$$
\begin{aligned}
{}^{12}\text{C} + p &\longrightarrow {}^{13}\text{N} + \gamma, \\
{}^{13}\text{N} &\longrightarrow {}^{13}\text{C} + e^+ + \nu, \\
{}^{13}\text{C} + p &\longrightarrow {}^{14}\text{N} + \gamma, \\
{}^{14}\text{N} + p &\longrightarrow {}^{15}\text{O} + \gamma, \\
{}^{15}\text{O} &\longrightarrow {}^{15}\text{N} + e^+ + \nu, \\
{}^{15}\text{N} + p &\longrightarrow {}^{16}\text{O}^*, \\
{}^{16}\text{O}^* &\longrightarrow {}^{12}\text{C} + {}^4\text{He}.
\end{aligned}
$$

The emission of a helium nucleus in the final stage regenerates the initial carbon-12. The latter thus plays the role of a catalyst. The overall result is the fusion of four protons into a helium nucleus. At high temperatures, this cycle dominates over the proton–proton chain. Indeed thermal agitation facilitates penetration of the relatively high electrical barrier between proton and carbon nucleus. Whatever hydrogen fusion mechanism is prevalent, the star's mass determines the rate at which it consumes its nuclear fuel, and hence also its lifetime. The higher its mass, the more quickly it burns.

The life of the Sun

Does the Sun empty itself of light? Our Sun is simply a star, in fact the closest of all the stars in the sky. Its brightness is due to its proximity (150 million km) and high surface temperature (almost 6000 K). An enormous sphere of gas in hydrostatic equilibrium, its stability is conditioned by the balance between gravity and the thermal pressure gradient. Each point of the Sun, torn between flight and fall, hangs suspended. Here nature has made one of its most wonderful equilibria. But it will not last forever, for the Sun shines, and to shine means death. It shines because it burns, and if it burns, it must be perishable. On the other hand, its life will be long, because in a certain sense it holds itself back from shining too much. This is the rule observed by a perfect star: to shine, but not too much.

The solar gas is extremely opaque, just like the early Universe which has reluctantly allowed a relic of its former brightness to filter down to us. The Sun, too, through its photosphere, has consented to release its light, but parsimoniously. This great light source is in fact almost completely opaque to its own brilliance. Light filters up from the depths, softening in its ascent towards the surface. The nuclear power supply buried deep below the solar mantle maintains the Sun's sheen and the heat of its substance. Its internal energy source arises from the atomic nucleus.

The Sun is a gravitationally confined nuclear reactor operating by thermonuclear fusion. This is the no-nonsense way we view our star today!

Its life can be summed up as a long battle between the nuclear fire and gravity. Its energy output is regulated by its opacity, that is, its ability to hold back or let through light, and not by the nuclear reactions themselves. The latter adjust so as to replace the radiated energy. To understand this, picture a funnel: the outflow is entirely controlled by the diameter of its lower opening. Energy escapes from the photosphere at a rate imposed by the opacity of the various layers that have to be crossed. This opacity depends on the composition, temperature and density of the layer in question. Resulting from the absorption and reemission of light by ions and (free or bound) electrons in the medium, this is the critical factor limiting energy losses.

All these basic principles and common sense arguments can be encoded into the mathematical language of physics. The equations that result can then be solved by computers. When applied to the Sun, the equations of physics bring off something quite miraculous: they make it transparent. The real Sun is transparent to neutrinos, whilst the simulated Sun is transparent to reason. We may read it like an open book. All the chapters concerning its inner workings, including its changes of colour and other surface features, are spelt out in our enlightened computer printout.

The star in the numerical model has an inside and an outside. The outside is defined as the limit beyond which it becomes transparent. This boundary is called the photosphere, or sphere of light, for it is here that the light that comes to us is finally emitted. It is thus the visible surface of the star, located at a certain distance R from the centre, which defines the radius and hence the size of the star. The photosphere has a certain temperature with which it is a simple matter to associate a colour, since to the first approximation it radiates as a blackbody, or perfect radiator. Indeed, the emissions from such a body depend only on its temperature. The correspondence between temperature and colour is simple. In fact, the relation between temperature and predominant wavelength (which itself codifies colour) is given by Wien's law, viz.

$$\text{wavelength (cm)} = 0.29 \times \text{absolute temperature (K)}.$$

Once the visible boundary of the model star has been established and the measurable global quantities worked out, the next step is to compare with the real Sun, if necessary fitting model parameters to obtain good agreement.

But what does the Sun provide us with in the way of measurable quantities? How can we guarantee the veracity of our model stars? The key quantities all concern the surface: intrinsic brightness, radius and effective temperature, that is, the temperature of the equivalent blackbody. But new constraints have now been added to this visible data, this time directly affecting the depths of our great star. These are the neutrino flux and global oscillations of the Sun's great body.

A saintly patience is required to obtain a good model of the Sun, and a diabolically proficient computer. Figure 5.2 shows an extract of past results obtained by the Saclay astrophysics department (France) for the case of a mature Sun. The whole exercise can be repeated for a Sun in its current condition, in its youth, or in its death throes.

The Sun is divided up theoretically into a large number of layers. The physical parameters, such as temperature, density, luminosity, energy production rate, emitted light and the rate of various nuclear reactions, vary with depth.

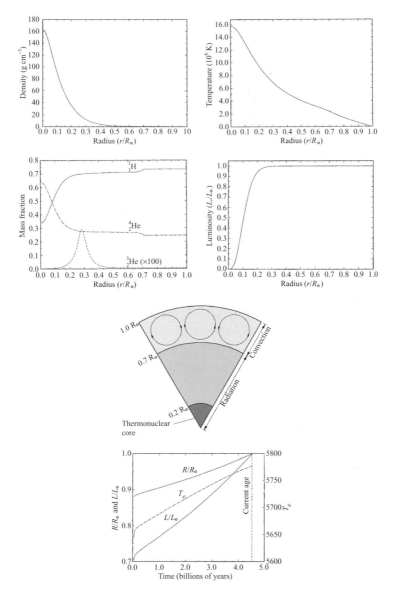

Fig. 5.2. Internal structure of the Sun. The top four graphs show the density, temperature, chemical composition and luminosity as a function of distance from the centre. Such profiles can be built up for each stage of the star's evolution. The figure shows the general regions of radiative transfer and convection in the bulk of the Sun. The bottom graph shows the gradual increase in radius, temperature and luminosity from birth.

The future of the Sun can be read off much better from the printout than from the tea leaves remaining at the bottom of a cup. The Sun's current characteristics, but also its past and its future, are all there in the printed matter spewed out by the computer.

We can reconstruct the whole career of the Sun, from its nebulous birth to its cloudy death, right up to its last breath, using a star model that covers all stars, spanning each of their layers as they are now and as they will become.

Nuclear core

The Sun shines by converting protons which it stores away in lavish quantities within helium nuclei. Six hundred million tonnes of hydrogen are burnt each second to maintain the sunshine we so delight in. Nuclear physicists have been toiling for half a century to work out the details of this transformation. However, it is only very recently that direct evidence has been brought to bear, which indeed tends to confirm the proposed mechanisms. The method used is rather novel. It consists in measuring the solar neutrino flux arriving on Earth, or to be more precise, underground. For neutrino astronomy is indeed an underground affair.

The results are globally positive, although a non-negligible discrepancy has come to light between prediction and observation. The latter is now thought to be due to certain ambiguities in our understanding of neutrino behaviour. However, this is not the place to enter into the subtle antics of these ghostly particles, which seem to have the singular habit of changing identity during flight. Suffice it to say that the mystery surrounding their versatile nature has proved very difficult to dispel. Moreover the physics of solar neutrinos has taken a new turn, very different from its original aims. The main objective is now to understand the neutrino. We have slipped imperceptibly from astrophysics to particle physics. But this saga has not ended and we may expect the balance to tip back the other way. The mystery of the solar neutrinos now belongs to the field of astroparticle physics.

Neutrinos from the Sun

The neutrino Sun never sets. Sixty billion neutrinos blasted out from the Sun's core eight minutes ago fly through every square centimetre of our body each second. We feel absolutely nothing and neither do they. They are the height of discretion. The night is not absolutely dark because, not only are we forever bathed in the cosmic background radiation at microwave wavelengths, but we

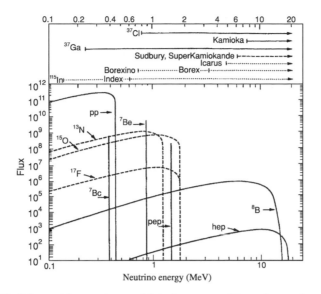

Fig. 5.3. Solar neutrino spectrum and detection ranges of the various underground neutrino observatories. The main part of the emission comes from the proton–proton reaction. About 3000 billion neutrinos pass through this figure every second. These neutrinos left the Sun eight minutes and two seconds earlier.

are also traversed from foot to head (and from head to foot during the day!) by flights of these invisible and almost intangible neutrinos.

Draw a square with side 1 cm and reflect on the fact that, every second, 60 billion neutrinos pass through it, neutrinos that left the Sun eight minutes and two seconds previously.

Each time a proton changes into a neutron in the Sun's core, a neutrino flies out and crosses the whole enormous body of the star as though there were nothing there. The Earth is a transparent ball for solar neutrinos and we are continually visited by these invisible beings.

Neutrino detectors are placed at great depths, at the bottom of mines and tunnels, in order to reduce interference induced by cosmic rays (Fig. 5.3). Two methods of detection have been used to date. The first is radiochemical. It involves the production by transmutation of a radioactive isotope that is easily detectable even in minute quantities. More precisely, the idea is that a certain element is transformed into another by a neutrino impact, should it occur. Inside the target nucleus, the elementary reaction is

$$\text{neutrino} + \text{proton} \longrightarrow \text{neutron} + \text{positron}.$$

The second method is based on detection of fast electrons induced by energy transfer from neutrinos to electrons in water. These fast electrons produce blue radiation, called Cerenkov radiation after the Russian physicist who discovered it. The latter is in turn detected by photosensitive cells or photomultipliers.

The first method brings into play enormous chlorine and gallium detectors. Seeing is in this case chemical, or more precisely, radiochemical. Strictly speaking, this technique is blind because it uses the transmutation of chlorine into argon and gallium into germanium, for example, without determining the direction of the incident neutrinos. We merely count events.

Historically, chlorine was the first target used to trap neutrinos. Chlorine-37 is mainly sensitive to high-energy neutrinos emanating from marginal fusion reactions (2 out of 10 000) which lead to production of boron-8. On rather rare occasions, under the impact of neutrinos, chlorine-37 is transformed into radioactive argon-37 which is easy to detect by its radiation. However, the myriads of low-energy neutrinos completely escape its notice.

Gallium can be used to detect low energy neutrinos arising in the reaction

$$\text{proton} + \text{proton} \longrightarrow \text{deuterium} + \text{positron} + \text{neutrino}.$$

Under the effects of the reaction

$$\text{neutrino} + \text{neutron} \longrightarrow \text{proton} + \text{electron},$$

gallium-71 transforms on very rare occasions into radioactive germanium-71. Much is gained by using gallium as target rather than chlorine because the far more numerous low-energy neutrinos are then accessible to measurement. However, the technological challenge is colossal.

Among electronic neutrino detectors is the great KAMIOKANDE experiment and its extension SUPERKAMIOKANDE. Spread out at the bottom of a mine in Japan, this device has directional sensitivity and it can thus be checked whether captured neutrinos do actually come from the Sun.

Five experiments have so far detected solar neutrinos. These are Homestake (USA), GALLEX, SAGE, KAMIOKANDE and SUPERKAMIOKANDE, all set up down mines or tunnels. Detected fluxes agree qualitatively with theoretical predictions, both in numbers and energies. We may say that we have basically understood how the Sun shines. The same set of nuclear reactions invoked to explain the solar luminosity does give rise to neutrinos.

Quite simply, each time a proton transforms into a neutron at the centre of the Sun, a neutrino flies out. Hence, neutrino experiments have definitely established, or so it seems, that the Sun shines because it carries out fusion of the nucleus of the simplest element, hydrogen, releasing energy

Table 5.1. *Solar neutrinos: confronting prediction with detection. The unit of flux is the SNU, or solar neutrino unit, which represents* 10^{-36} *neutrino events per target atom per second*

	Chlorine	SuperK	Gallium (Gallex)	Gallium (SAGE)
Experiment	2.55 ± 0.25	2.44 ± 0.26	76 ± 8	70 ± 8
Prediction	7.1 ± 1.7	5 ± 1.3	127 ± 8	127 ± 8

Source: Courtesy of Sylvaine Turck-Chièze, CEA, France.

according to Einstein's famous formula $E = mc^2$. The architects of the theory of fusion reactions in stars, Hans Bethe and Fred Hoyle, should rejoice. $E = mc^2$, and the Sun shines! Einstein would be rubbing his hands together with glee.

But quantitatively there is a disagreement between measurement and theory. The neutrinos actually caught in our traps are two or three times less numerous than the theory says they ought to be (Table 5.1). Naturally, this discrepancy is a challenge to the physicist. It does not seem to arise from any problem with the solar model itself, now that the Sun has been probed by its mechanical oscillations.

The tendency today is to incriminate the cunning neutrino, with its three faces, which escapes the vigilance of radiochemical detectors because it wears a disguise. More seriously, it may be that neutrinos, given their very small mass, are able to oscillate between different states, of which only one can be detected. In this case, a large part of the neutrino flux would escape detection. This is not the place to go into further detail as regards this intricate problem. However, a whole range of experiments, including the SNO (Solar Neutrino Observatory) at Sudbury in Canada with its thousand tonnes of heavy water, should be able to reveal the true nature of these multifaceted particles. Indeed, recent results from SNO and SUPERKAMIOKANDE combine to confirm the hypothesis of neutrino oscillation, a pure quantum process made possible by a mass difference between neutrino species. (For recent measurements and their interpretation, see Bahcall (2002).)[1]

---

[1] I would like to pay tribute to Saclay's inveterate neutrino tasters, Michel Spiro, Daniel Vignaud and Michel Cribier, and also to Sylvaine Turck-Chièze and her students who prepare the solar sauce.

Stellar model

Stars are suns of varying masses, at different stages in their evolution, as our models will show. Some like to design protoype cars or aeroplanes, others wedding dresses, whilst we astral physicists make evolutionary models of stars.

We are thus able to reconstruct the Sun's whole career, from its nebulous birth, through its first nuclear reaction, right up to its last breath as a red giant. We may watch it swell up proudly as it burns its first helium, then throw its envelope off a hot, dense core, before the latter grows rigid, becoming a white dwarf, frozen in crystal. Wonders! But let us recall the vast effort that went into this undertaking.

The mathematical architecture of the model had to be built up from scratch. The structure equations were put together like the timbers of a house and filled out with physical data describing the behaviour of matter at high temperatures, where atomic and nuclear physics come into play.

The numerical star sculptors had to grapple with transparency and opacity, the origin of solar light and the propagation of radiation inside the Sun. For example, nuclear physicists were called in to provide the probabilities of the various nuclear reactions that occur inside our star, probabilities that depend on the nuclei present and the temperatures prevailing at different depths. These nuclear physicists brought all their science to bear, working out reaction rates by studying at close quarters collisions between accelerated nuclei and appropriately chosen targets. The particle accelerator has thus become a tool for astrophysics.

Finally, the numerical model had to converge, no mean achievement from the computing point of view. It is with obvious satisfaction that one sees the results emerge at the end of the day, neatly arranged on the computer printout.

The model Sun tells all. We may read off its temperature, density, chemical composition, luminosity and nuclear reaction rates at any depth and any stage of its evolution, from the youthful Sun, to its current middle age and forthcoming old age. The day-star has become limpid and with it every other star.

Naturally, a good model does not merely describe what is known, restricting itself to the purely apparent, or one might say just keeping up appearances. That is, it is not concerned merely to reproduce the outward features of the star, like size, luminosity and colour, but also to reveal its invisible inner workings down in the vertiginous depths. It can thus help to predict certain unknown aspects, calling for new types of observation, such as attempts to stop a few neutrinos in their tracks among the hordes escaping incessantly from its core.

The model exposes the invisible truth, hidden since the Sun first came into being: its core contains a nuclear power generator. In addition, it predicts the future of our star and all stars of comparable mass: the Sun will become a red giant, its envelope will fly off to leave a dense, white core.

The Sun is currently in the simplest and longest-lasting phase of stellar evolution, the quasi-static phase of core hydrogen burning, also known as the main sequence by astronomers who untiringly refer to Hertzsprung and Russell's diagram with temperature on the abscissa and luminosity on the ordinate. The temperature at the centre of the Sun is currently 15 million K. Of course, this temperature cannot be measured directly, but it can be precisely calculated from a physical model of the star, itself tested by neutrino flux measurements and observations of solar oscillation modes that tend to corroborate it. But unfortunately, when dealing with other stars, we do not have these figures and the HR diagram is the only tool at our disposal to pin down our theoretical models. Numerical simulations must be continually compared with real stars, at least, with their outer appearances, that is, their surface radiation and temperature.

As we cannot get hold of the stars, the galaxies and other universes, as we cannot manipulate them, we make models. In this way, numerical experiments provide support for telescopic observations, and for real experiments, such as those carried out in particle accelerators, which simulate energy conditions in the Big Bang and stars.

Today, particle accelerators and computers are as much a part of astronomy as telescopes intent on spying out the visible and the invisible. In their accelerators, high-energy physicists are able to reproduce conditions in the Big Bang and in the stellar core. Then, taking over from them, numerical simulation by computer can write the story of matter through its various cycles of concentration, nucleosynthesis and dispersion.

These models describe not only the Sun in all its layers as they are today and as they will be tomorrow, but the same for every other star. The future of the chosen star is revealed in a temporal sequence of steps and the same goes for the galaxies and even the whole Universe.

The story of matter can be read in the evolutionary models fashioned by physicists, astronomers and mathematicians working in unison. Models of stars allow us to determine their mass and their age on the basis of their colour and brightness. But these paper stars, for all their equations, must forever be confronted with real ones.

Shouts and whispers

Constancy and majesty, but also fury and destruction, the whole range of emotions is depicted there, in the sky. Beneath the serene countenance displayed to human inspection is dissembled an inner, invisible fire and wrath. The reassuring placidity gives from time to time a hint of the storm in its heart. How out-of-date is the comparison between the perfect order of the planets as they move steadily across the zodiac and the chaos that typifies human life!

Supernovas, quasars, black holes, raging galactic nuclei, these are what now dominate astronomical thought. And the cosmic menagerie grows richer from night to night. Today, no science needs updating so regularly. Celestial bodies appear, accomplish their task and disappear into the apparent blackness of space.

The deceptive serenity of the sky as it has caressed the eye of humanity from time immemorial is perhaps the very condition for the nurture of rational thought. If from the beginning our eye had seen all these phenomena, and not only the calm suns, would our soul not have suffered perpetual torment? In the last century, was the revelation of creative violence that unendingly shakes and reshapes the Cosmos the prelude to a spiritual transformation of our civilisation? In the words of Roberto Matta, the great surrealist painter: revolution means changing the sky.

The revelatory power of the new astronomy, especially astronomy associated with the extreme forms of radiation, resides in its capacity to expose previously unknown processes to reason and understanding: gamma astronomy, the most violent phenomena in the Universe, such as the rupture and destruction of stars, and infrared astronomy, the gentle events, such as the birth of stars. Optical astronomy fills the relatively calm gap between stellar birth and death, whilst millimetre radioastronomy opens our minds to the formation of molecular structure in great clouds of cold gases and opaque dusts, far from any devastating light.

From violent beginnings to a final regenerating apotheosis in explosion, originating in a gigantic deflagration, matter is brewed, torn apart, divided, ground, ejected, scorched, now and for evermore by the stellar monsters. Taking refuge in cold clouds, it aggregates into molecules and tiny grains that filter starlight. Violence here, calm and stability elsewhere.

As the new sky takes shape, it remains to give meaning to it. The Universe cannot be conceived entirely on the basis of a model of its content. Our task is to rediscover some form of order, and this order must be temporal. The Universe is the product of its history, as well as the sum of its atoms.

Many learned men and women would say that the story told by astronomers is no less extraordinary than the most magnificent legends, the epic of Gilgamesh or the Bible, and I am inclined to agree. The astronomer's sky is no less opulent than that of the prophet or the poet, itself replete with flying arrows and burning celestial passions. But it has every chance of being more realistic, for the sky is a genuine treatise on stellar alchemy and the genealogy of matter.

The story of the stars, bright and dark

The story of the Sun, and indeed of any star, centres around the age-old struggle between the nuclear furnace and gravity. The pressure due to heat, or rather heat gradient, fights back against collapse and a masterful equilibrium is achieved. The vocation of the perfect star is to burn, in the nuclear sense, that is, to transmute elements in nuclear reactions.

But the description of the stars' accomplishment would be incomplete if we did not mention a more secret process of densification and unification of matter which goes hand in hand with nucleosynthesis and the dissemination of the wrought matter. The latter diversifies nuclear species and disperses them, whilst the former imposes uniformity and traps matter in a gravitational prison, a kind of stellar corpse.

The stars are also involved in moonlighting. Heavenly hourglasses, at the final moment they unveil the transcendent pearl, white dwarf, neutron star or black hole, hatched secretly in the deepest vault, whilst the envelope flies with the wind, loaded up like a galleon with all their wealth and splendour, a largesse of fertile ash by the name of carbon, nitrogen and oxygen. They do not know what they are doing. They are pouring wagonloads of future flesh into the sky!

Stars drive galactic evolution. At the end of their existence, they inseminate space with the products of their nuclear alchemy. Then, in dark clouds, sheltered from ravaging photons, molecules build up. Stars and planets are constantly being born in the cold of space.

The words 'birth', 'life' and 'death' are commonly used in popular astronomy books to describe the cycle of the stars. However, let us say at the outset that stars do not fulfil the criterion of living matter, even though their appearance changes as they evolve and their sudden disappearance can sometimes resemble the brutality of death, at least for those in the supernova category. By definition, a living being continually draws energy from its surroundings, whilst a star is self-sufficient, obtaining energy from its own substance.

A star goes out like a burning log when it has used up all its combustible component, and it shines only because it transmutes the chemical elements. It is not a being but a state. A visible star is merely a fire, of nuclear constitution. The twinkling star is the bright phase in the cycle of matter, and then not of all matter, since neutrinos and other particles that are not susceptible to strong and electromagnetic interactions, if such particles should exist, do not play a part in structuring matter into objects. Moreover, there are invisible bodies like black dwarfs, neutron stars and black holes, which are the direct descendants of visible stars.

Stars are not living bodies, but objects that change structure and appearance. They evolve and proliferate, and that is why they are said to be born, to live and to die.

At the end of its evolution, the star restores part of itself, containing the ashes of its past activities, to its surroundings. The sky gains in heavy elements, as earth grows richer in salts. For this reason, the medium separating stars also evolves. When the heavy element content reaches 2%, life and consciousness emerge, as we may deduce from the only known case: our own.

Massive stars are rightly considered to drive galactic evolution, both in the chemical sense and in the mechanical sense. They emit great quantities of UV light because of their high surface temperatures, and this ionises the neighbouring galactic medium. Interstellar dust heated by UV radiation from massive stars reradiates the received energy in infrared form. These stars scald and stir the interstellar medium, injecting the new atoms they have fashioned. They accelerate particles, so-called cosmic rays, to a significant fraction of the speed of light. These shatter atomic nuclei of carbon and oxygen deposited along their path by defunct stars, thereby producing the light and rare species lithium, beryllium and boron. Lithium would be used much later to treat the nervous breakdowns of astronomers trying to work out where it came from. Boron, incorporated into borosilicates, inhabits many ovens in the form of Pyrex cooking dishes. Beryllium bestows its intense colour on seaweed and sapphire.

For those who would like to construct a similar universe on paper, an adjustment would be needed in one direction or another to avoid its being completely devoid of complex nuclei. A universe made up entirely of hydrogen would not be very different from ours, but if all stars had originally been composed of helium, they would have burnt out very quickly. A helium Sun would have burnt for only about 10 million years. The planets would very probably not have had time to form. There would be no water.

At the present time, the stars form a crowd of suns of different masses at different stages in their evolution. They burn their fuel, mainly hydrogen, at a leisurely pace, then turn to helium, the ash of the first reaction, followed by carbon, oxygen and finally silicon, provided that their core temperature (mass) is sufficient. Silicon combustion leads to iron which will not burn in the nuclear meaning of the word, for it is the most stable nucleus known to nature. From this moment, the star is doomed. The long struggle between the nuclear fire and gravity begins to turn in gravity's favour. The core collapses, and the outer layers drop down and bounce back up again. Explosion follows implosion a split second later. The star showers space with the atoms it has made. Hence, the most fertile objects in the sky are the supernovas (see Appendices 3 and 4).

All the work involved in astrophysical reflection aims to give meaning to the word 'star'. It turns out that the description of stellar evolution complies with known physical theories and nuclear processes.

Source of atoms

Everywhere in the Universe the same patterns reappear: atoms in the microcosm and stars in the macrocosm. Atoms and stars are related, for the star is the mother of atoms. This statement sums up one of the great discoveries of the twentieth century. The task of nuclear astrophysics is to draw in the details of the relationship. The creation of matter has become a subject of scientific research. The source of most atoms, or rather their nuclei, is the stars.

But what is a star? An artisan, a bee, a forge, a physical system that operates at very high pressures? In the general economy of the Universe, stars play the role of conscientious craftsmen. They are the site of nuclear alchemy. Matter partly dematerialises to make light, whilst the simple complexifies: hydrogen transmutes into helium, helium into carbon and oxygen, oxygen into silicon, and silicon into iron.

The star burns its own ashes, and the ashes of its ashes, but iron will not burn and the nuclear fire goes out. Blue giants with collapsed cores open like flowers and scatter their swarms of winged atoms in the sky. The dying of their light is acknowledged with the cry of 'Supernova!'

The parent stars of these supernovas are blue, precisely because they are hot and massive. Generation after generation of blue stars drive the ecology of the galaxies, enhancing the beautiful precipitated clouds of interstellar compost with elements so essential to life: carbon, nitrogen, oxygen, magnesium, silicon, sulphur and the rest. Remember then our blue ancestors, mother of our atoms; it is time we recognised them, and their own nebulous ancestors, light, and its father, the prospering void.

The theory of the Big Bang, natural child of general relativity and astronomy, has conjoined with nuclear physics to teach us that hydrogen and helium come down to us from the very beginning, and that the stars in the guise of gravitationally confined nuclear reactors have manufactured all the other elements from carbon to uranium in their crucible from simpler elements. They have opened like flowers, letting fly myriad atoms of every variety, and thus sowing the seeds of life. The unborn are there, around the supernova, a torn star that has relinquished its substance. The link between stars and humans is thus material, genetic and historical.

The study of the synthesis of atomic nuclei via nuclear reactions which proliferate in certain hot objects or violent astronomical events may well appear a superhuman task. It is true that it has not been easy to understand the synthesis of the chemical elements and the confection of metals in the stars. But patience and determination have eventually dispelled the more resistant mysteries and in recent years significant progress has been made in understanding the origin of the elements. The main sites of nucleosynthesis have now been assigned to the various species of atomic nucleus.

The planets do not feature amongst these sites. It has become totally unreasonable to attribute the paternity of quicksilver (mercury) to the planet Mercury, to associate iron with the red planet Mars, and lead with Saturn. We know today that iron, lead and mercury come from supernovas.

Synopsis of nucleosynthesis

Like chemistry, nuclear astrophysics is a combinatorial art. Nuclear reactions are written down like chemical reactions, replacing atoms with nuclei.

The earliest nucleosynthesis belongs to cosmology. A first flurry of nuclear reactions took place in the great primordial heat occurring between 1 and 100 seconds after the Big Bang. The species H, D, ^{3}He, ^{4}He and ^{7}Li were synthesised in proportions that depend on the baryonic density of the Universe (Vangioni-Flam & Cassé 1998) (see Appendix 1).

This sequence effectively breaks off at atomic mass 7 because the heirs of helium are so incurably unstable. In particular,

$$^4\text{He} + {}^4\text{He} \longrightarrow {}^8\text{Be (unstable)}.$$

The Universe had to make stars in order to pursue the long road to nuclear complexity.

Lithium, beryllium and boron are products of nuclear spallation, that is, the shattering of heavy nuclei, mainly carbon, nitrogen and oxygen (collectively denoted CNO), themselves produced in stars. Fixed or moving CNO nuclei fragment upon collision with fixed or moving H and He to form the light and fragile species ^{6}Li, ^{7}Li, ^{9}Be, ^{10}B and ^{11}B. In fact they are so delicate that they are destroyed in stars.

In a classic article published in the journal *Review of Modern Physics* in 1957, Burbidge, Burbidge, Fowler and Hoyle described the various processes responsible for synthesising chemical elements during the evolution of stars. These processes include thermonuclear fusion and neutron capture.

Iron is the chemical element with the highest binding energy in the nucleus. Consequently, the thermonuclear fusion process cannot proceed beyond this point. The successive combustion stages (hydrogen, helium, carbon, neon, oxygen and silicon) require ever-higher temperatures to overcome the electrical repulsion between reacting nuclei. The ashes from one cycle serve as fuel for the next. Hence, carbon is the ash of helium, which is itself the ash of hydrogen. Hydrogen is the best fuel in terms of energy production and the following cycles yield less and less energy. This also means that they last for shorter and shorter periods.

The nature and duration of combustion phases depend on the mass of the star in question. Only stars with masses greater than $8\,M_\odot$ can string together all the fusion cycles.

Remaining for a moment with the massive stars, a distinction must be made between slow, secular, quasi-static and explosive nucleosynthesis. The time-scale in the latter case is of the order of 1 second and it only affects the innermost layers of stars, rich in silicon, oxygen and carbon.

Beyond iron, nucleosynthesis proceeds via neutron capture by iron and its neighbours. Two types of neutron capture, slow denoted by s and rapid denoted by r, come into play depending on the intensity and duration of neutron irradiation. Once the neutron has been absorbed, the resulting product depends on whether the neutron has time to convert into a proton inside the nucleus before a further neutron is absorbed. If the transmutation occurs before further capture, we have an s process, otherwise an r process.

In the case of rapid capture, several neutrons are added before conversions of type $n \rightarrow p$ bring the neutron to proton ratio back to reasonable proportions. The r process requires impressive neutron fluxes and extreme densities and temperatures that can only be achieved in type II supernovas or the coalescence of two neutron stars. The details are not yet understood. However, we have no other explanation for the existence of gold and heavy isotopes of tin (^{121}Sn and ^{124}Sn), for example. There is another process, namely photodisintegration, which is very short-lasting and leads to nuclei poor in neutrons, or rich in protons (referred to as a p process).

In the slow process, neutron captures are separated by $n \rightarrow p$ conversions, so that the result never lies far from the valley of stability. The s process takes place in much less extreme conditions. There are two forms.

1. The main s process synthesises neutron-rich nuclei with atomic number A greater than 100 and occurs in asymptotic giant branch (AGB)-type red giant stars undergoing thermal palpitations.
2. The weak s process produces nuclei between $A = 60$ and 100 and accompanies helium fusion in massive stars.

By coupling the nuclear reaction network to stellar models, we may calculate the compositions resulting from these nuclear processes, under any imaginable set of conditions. Reaction probabilities are supplied by nuclear physicists. Thanks to these stellar numerical models armed with the relevant microscopic data, the star becomes transparent and we may follow its career right through from nebulous birth to lustrous death, watching the nuclear combustion cycles string together.

Hydrogen fusion via either the proton–proton chain or the CNO cycle in the centre of stars comes to an end when most of the hydrogen has been transformed into helium. Helium fusion produces two elements essential to life, namely carbon and oxygen. In fact, carbon constitutes 18% of our bodies, and oxygen 65%, whilst the fractions of these same elements in solar material are just 0.39% and 0.85%, respectively. Only hydrogen and helium are more abundant in the Sun.

Theoretically, helium fusion is inhibited by the absence of stable nuclei with masses between 5 and 8, which would bridge the gap to carbon. However, the existence of carbon-containing structures such as ourselves implies that the stars have managed to get around this difficulty. Unable to produce carbon by two-body reactions, they build from three ingredients: $3\,^{4}\mathrm{He} \rightarrow\,^{12}\mathrm{C}$. This triple reaction encounters very favourable conditions in red giants, with core temperatures of 100 million K, because an excited state of carbon exists there (at 7.65 MeV). This excited state considerably increases the probability of reaction. Without this providential energy level, predicted by Hoyle in 1954 and discovered later, carbon could not have been produced by red giants and life would have gone another way. Let us say no more.

The reaction $\alpha +\,^{12}\mathrm{C} \rightarrow\,^{16}\mathrm{O} + \gamma$ converts a substantial fraction of the carbon into oxygen. Oxygen production is sometimes followed by neon production via the reaction $\alpha +\,^{16}\mathrm{O} \rightarrow\,^{20}\mathrm{Ne} + \gamma$.

In stars of mass less than $8\,M_{\odot}$, core fusion then breaks off, leaving an inert star whose core of carbon and oxygen is degenerate in the quantum sense of the term. This is the white dwarf. The word 'degenerate' comes from the vocabulary of quantum theory, which in this case applies to electrons in conditions of very high density and relatively low temperature. By virtue of the Pauli exclusion principle, the electrons refuse further compression and indeed exert a pressure which opposes gravitational contraction of the star. The pressure is no longer proportional to the temperature. Matter has lost its gaseous flexibility.

Beneath their placid exterior, white dwarfs are very sensitive to the addition of matter from without. The slightest excess ends in a brilliant flash of light (a nova) or an explosion (a type Ia supernova).

Another important reaction chain for future phases, including the final nuclear statistical equilibrium (see below), is one which enhances the neutron content of matter via the weak interaction (β^+):

$$CNO \longrightarrow {}^{14}N,$$
$$^{14}N + \alpha \longrightarrow {}^{18}F + \gamma,$$
$$^{18}F \longrightarrow {}^{18}O + \nu + e^+,$$
$$\alpha + {}^{18}O \longrightarrow {}^{22}Ne + \gamma,$$
$$\alpha + {}^{22}Ne \longrightarrow {}^{25}Mg + n.$$

Helium fusion moves directly from helium to carbon, leaping across the lithium–beryllium–boron trio. These nuclei are not produced in stars. Indeed, they are destroyed there, as a result of their excessive fragility. They are generated in the interstellar medium by collisions between high energy nuclei and protons and helium nuclei at rest, and also by the opposite process which amounts to swapping over target and projectile, as already mentioned.

In stars of mass greater than 8 $M_\odot$, thermonuclear fusion can proceed further. Their cores continue to heat up. The evolution is greatly accelerated by emission of neutrinos and antineutrinos. The characteristic time-scale of nuclear evolution in the core becomes shorter than the time required by the visible envelope to readjust. It follows that the internal evolution does not result in any changes in colour or luminosity. Everything happens as though a star made up of heavy elements were living and evolving within a red supergiant (in the lighter cases) or a blue supergiant (for the more massive stars).

When helium fusion comes to an end at the centre of the star, its carbon/oxygen core contracts and the temperature rises. At a temperature slightly below 1 billion K, carbon is ignited.

Carbon fusion produces neon, sodium and magnesium via reactions of the following type:

$$^{12}C + {}^{12}C \longrightarrow \begin{cases} ^{24}Mg + \gamma, \\ ^{23}Mg + n, \\ ^{23}Na + p, \\ ^{20}Ne + \alpha, \end{cases}$$

where γ, n, p and α denote a gamma photon, a neutron, a proton and a helium nucleus, respectively.

Neon fusion takes over when the temperature reaches 1 billion K. At this temperature, thermal photons have enough energy to knock fragments off neon nuclei in the form of helium:

$$\gamma + {}^{20}Ne \longrightarrow {}^{16}O + \alpha.$$

Helium nuclei released in this way are captured by as yet unbroken neon nuclei to form magnesium-24:

$$\alpha + {}^{20}\text{Ne} \longrightarrow {}^{24}\text{Mg} + \gamma.$$

When neon combustion is concluded, the star comprises mainly oxygen and magnesium.

Oxygen fusion begins when the temperature reaches some 2 billion K. The main reactions are:

$$^{16}\text{O} + {}^{16}\text{O} \longrightarrow \begin{cases} {}^{32}\text{S} + \gamma, \\ {}^{31}\text{P} + \text{p}, \\ {}^{31}\text{S} + \text{n}, \\ {}^{28}\text{Si} + \alpha. \end{cases}$$

The most important reaction product here is silicon-28, which results from the addition of two oxygen nuclei accompanied by the loss of one helium nucleus.

In the case of silicon fusion, which begins at around 2 billion K, the reactions proceed in a slightly different manner and we return to a fusion scheme similar to that of neon. At this temperature, silicon nuclei are gradually gnawed down by thermal photons which detach helium nuclei, protons and neutrons from them. These light nuclei combine with intact silicon to give nuclei in the region of iron. Schematically,

$$^{28}\text{Si} + \gamma \longrightarrow 7\alpha,$$
$$^{28}\text{Si} + 7\alpha \longrightarrow {}^{56}\text{Ni},$$
$$^{28}\text{Si} + \gamma + \text{p} + \text{n} \longrightarrow \text{iron peak}.$$

These reactions, which are extremely fast, are often balanced by the reverse reactions, so that an approximation known as nuclear statistical equilibrium can be applied. In this case, the most stable species, i.e. those possessing the highest binding energy, are favoured. The result depends on only three parameters, viz. temperature, density and neutron/proton ratio. The latter in its turn results from the previous nuclear reactions and the composition of the star at birth, through neon-22 (see above).

Carbon and neon fusion on the one hand and oxygen and silicon fusion on the other are grouped together because they very often lead to similar results.

In the final stages of their evolution, stars take on an onion-like structure, with a central iron core and successive shells of silicon, neon, carbon and oxygen, helium and hydrogen. At this stage the core is dominated by iron-54 and is degenerate because of its high density, whilst above it silicon, oxygen, helium and hydrogen burn in their respective layers. The star shines by the light of its last remaining embers.

The following stage is core collapse caused by electron capture or photodisintegration of iron. According to the traditional view, collapse leads to formation of a neutron star which cools by neutrino emission and decompression of matter when it reaches nuclear density (10^{14} g cm^{-3}). The rebound that follows generates a shock wave which is capable of reigniting a good few nuclear reactions as it moves back out across the stellar envelope.

Strengthened by absorption of high-energy neutrinos, the shock wave expels the outer layers. Seen from outside, the star appears to have exploded. On Earth, this fabulous event is greeted with the joyous shout of 'Supernova!'

The sudden heating caused by the shock wave and the rapid cooling that ensues therefore rekindle nucleosynthesis. The composition of the silicon shell and part of the oxygen shell is modified by explosive nucleosynthesis, although it only lasts for a split second. This differs from previous forms of nucleosynthesis in that the weak interaction is as though frozen, or paralysed. It is the radioactive nuclei produced in this brief event, mainly nickel-56, that make the supernova shine with diminishing light curve as time goes by (Fig. 5.4).

Globally, then, oxygen, neon and magnesium originate in hydrostatic shell combustion and the quantity synthesised and ejected increases with the mass of the progenitor star, whilst sulphur, argon, calcium and iron are essentially due to explosive nucleosynthesis and the ejected mass is much less variable from one star to another.

In this way we climb the stairway to nuclear complexity.

S- and r-process isotopes

Elements beyond iron cannot be efficiently produced by nuclear fusion because of the strong electrical repulsion occasioned by their highly charged nuclei. The temperatures required to overcome this repulsion are so high that no nucleus can survive the subsequent bombardment by rampaging photons and the resulting photodisintegration. Even iron is destroyed. For this reason, heavy elements can only be synthesised by successive neutron captures on elements around the iron peak, interspersed by transmutation of neutrons into protons (β decay) within the resulting nuclei.

Neutron capture may occur over a time-scale that is long enough to allow all possible β decays to take place, in which case we refer to the s process, or quite the opposite may happen, and we have the r process, as discussed earlier. The two processes lead to quite distinct abundance distributions (Fig. 5.5).

In a continuous flow of neutrons, the abundance of each element is inversely proportional to the probability (cross-section) of neutron capture. Nuclei

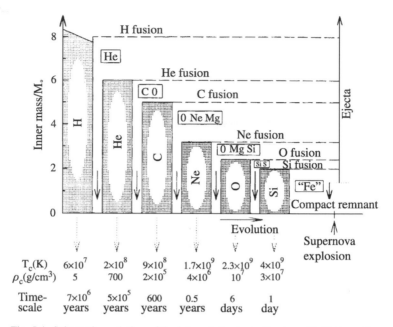

Fig. 5.4. Schematic evolution of the internal structure of a star with 25 times the mass of the Sun. The figure shows the various combustion phases (shaded) and their main products. Between two combustion phases, the stellar core contracts and the central temperature rises. Combustion phases grow ever shorter. Before the explosion, the star has assumed a shell-like structure. The centre is occupied by iron and the outer layer by hydrogen, whilst intermediate elements are located between them. Collapse followed by rebound from the core generates a shock wave that reignites nuclear reactions in the depths and propels the layers it traverses out into space. The collapsed core cools by neutrino emission to become a neutron star or even a black hole. Most of the gravitational energy liberated by implosion of the core (some 10^{53} erg) is released in about 10 seconds in the form of neutrinos. (Courtesy of Marcel Arnould, Université Libre, Brussels.)

with closed neutron shells ($N = 50, 82, 126$) oppose the adjunction of further neutrons and this leads to accumulation and abundance peaks for these numbers.

Likewise, even nuclei (with even numbers of protons) have lower probability (cross-section) for neutron capture than odd nuclei, and this results in a greater abundance of the former. The even–odd imbalance is manifested once again.

The s process builds up an abundance distribution with peaks at mass numbers ($A = Z + N$) 87, 138 and 208 and pronounced even–odd imbalance. The main component of the s process is associated with thermal pulsations of stars in the asymptotic giant branch (1–3 $M_\odot$) which produce neutron densities between 10^7 and 10^9 cm^{-3} (Fig. 5.6).

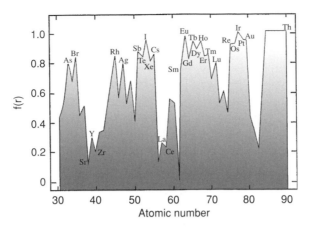

Fig. 5.5. Decomposition of Solar System abundances into r and s processes. Once an isotopic abundance table has been established for the Solar System, the nuclei are then very carefully separated into two groups: those produced by the r process and those produced by the s process. Isotope by isotope, the nuclei are sorted into their respective categories. In order to determine the relative contributions of the two processes to solar abundances, the s component is first extracted, being the more easily identified. Indeed, the product of the neutron capture cross-section with the abundance is approximately constant for all the elements in this class. The figure shows that europium, iridium and thorium come essentially from the r process, unlike strontium, zirconium, lanthanum and cerium, which originate mainly from the s process. Other elements have more mixed origins. (From Sneden 2001.)

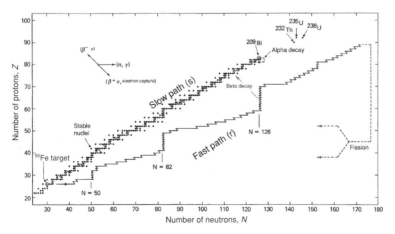

Fig. 5.6. Path of s and r processes across the (Z, N) plane. Everything begins with iron. The s process follows roughly along the valley of stability, flowing like a river along the banks it defines. It ends with the α decay of bismuth-209. The r process takes matter far out of the valley on the neutron-rich side, whilst the weak interaction brings it back to the fold. In this case neutron capture continues until the nucleus undergoes fission. The climb to neutron-rich summits is indeed vertiginous.

The distribution of r-process nuclei is characterised by peaks shifted to 80, 130 and 195, and an absence of the even–odd imbalance. There is still debate about where exactly the r process occurs, although supernovas have been suspected from the very beginning (Burbidge *et al.* 1957). According to the most popular model, it takes place in the hot (high entropy) bubble that surrounds incipient neutron stars. In this region the high photon/baryon ratio favours photodisintegration and this in turn generates a large number of neutrons and a small number of iron nuclei. Each iron nucleus can thus be frequented by a great many neutrons (10^{20} cm^{-3}). As they have no electrical charge, these neutrons may insidiously introduce themselves into the iron, thereby transforming it into some neutron-saturated exotic species. The latter then stabilises through a chain of β decays, transforming neutrons into protons within the nucleus. Although it seems attractive, this model nevertheless encounters problems and indeed opposition from some quarters.

6

Sociology of stars and clouds

Glossary

Bok globules dense, compact and cold interstellar clouds
Doppler effect frequency shift between emitted and received light due to the relative motion of the source and observer
lepton light particles of the same family as the electron
microquasar miniature quasars discovered in our own Galaxy
photodissociation breakup of molecules by photons
spark chamber device for detecting gamma rays by production of electron–positron pairs
superluminal velocity a velocity apparently greater than the velocity of light

Galaxies: structural units of the cosmos

On a beautiful night a faint band of diffuse light may be seen sloping across the sky from one horizon to the other. This non-uniformity, or anisotropy, in the distribution of stars is the stamp of our own Galaxy, a flattened disk containing billions of stars. We observe the disk edge-on from a point on the outskirts, exiled some two-thirds of the way along the galactic radius, about 30 000 light-years from the centre.

So we are not even at the hub of our own star system! The disk itself is only a few hundred light-years thick, if we exclude the central bulge and halo, a spherical region filled with ageing stars.

The mean distance between any two stars is of the order of a few light-years. Even the nearest stars are so far away that we perceive them as mere points of light. Amongst all the stars, the Sun located just 8 light-minutes away is the only one close enough to appear as a disk. It is an incandescent sphere of gas

105

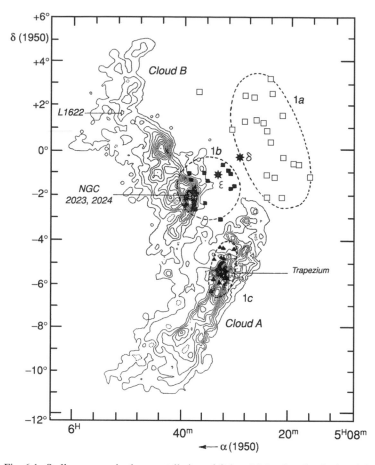

Fig. 6.1. Stellar nursery in the constellation of Orion. Molecular clouds *A* and *B*
were detected by their radio emissions. They appear to have given birth to several
generations of stars (1*a*, 1*b* and 1*c*).

over 1 million km in diameter and its brightness and warmth are entirely due
to its proximity.

Immense clouds of cold and rarefied gases float between and around the stars.
These clouds are mainly composed of hydrogen, the most widespread element
in nature. Stars form there and contribute their ashes to them when they reach
the end of their evolution (Fig. 6.1).

The flattening of the Galaxy suggests that it is rotating about an axis per-
pendicular to the disk, and this is indeed confirmed by direct observation of
the large-scale motions of the stars. In fact, they orbit about the galactic centre,
making a complete revolution every 200 million years.

The visible Universe is populated by a whole range of galaxies of different shapes – spiral, elliptical or irregular. Their size also varies, from a few thousand to a few hundred thousand light-years. They occur alone or gathered into colossal groups. Distances separating galaxies are of the order of 1 million light-years. They are all rotating, and those spinning most rapidly are spirals. Irregular galaxies contain large quantities of gas, the spirals less and the ellipticals almost none at all.

Apart from rotating about their axes, the galaxies display systematic motions relative to one another. In fact, they are moving apart at speeds proportional to the distance between them. The recession speed amounts to some 100 km s^{-1} for every 3 million light-years of separation. This overall motion is the clearest evidence we have for the expansion of the Universe.

How did the galaxies and protogalaxies form? How did matter gather up in this way, despite the expansion of the Universe which tends rather to dilute it?

Since distances grow as time goes by, the galaxies were necessarily closer together in the past. Was there a time when they were actually fused together into one? Was this the origin of the Universe, the beginning of time? Not at all! Neither stars nor galaxies could have existed in the primordial furnace, for if we reverse the arrow of time, the Universe grows hotter. And when the sky is hotter than the stars, those stars must dissolve into the sky. When the temperature of the sky background was several thousand degrees (equivalent to the surface temperature of a typical star), matter would not have been able to collect together into stars. In fact, stars were made to shine in a black and cold sky. They are designed to give light, not to receive it. Consequently, stars and galaxies could not have existed as such in the early Universe, when light and matter were compounded.

As a rule, cosmology concerns itself with the evolution of the Universe and is more transparent than cosmogony, the study of creation. Births, of the Universe, galaxies or stars, or originating events of any kind represent the more obscure episodes in the history of the Cosmos. It is no surprise to observe that the notion of 'origin' is more opaque than the notion of 'evolution'. One day we will be able to observe the primary genesis of the architectural themes of our Universe. There are signs that the birth of galaxies will soon be witnessed, and we already have a theoretical premonition.

Only those scraps of matter more concentrated than the medium in which they have accumulated can foretell the emergence of cosmic forms. And these embryonic galaxies can only be random concentrations of the undifferentiated magma comprising photons, baryons and non-baryonic particles.

Basic models of massive non-baryonic matter (of neutralino type) suggest a hierarchical formation of structures. They work on the hypothesis that primordial density fluctuations develop in the dark substrate, and that they grow and prosper under the effects of gravity. In this way small structures made up of both baryonic and dark matter are the first to condense out. The baryonic clouds cool down and lose enough kinetic energy to come under the sway of their own gravity. Collapse by condensation of particularly dense and cold regions is then likely to bring about the formation of stars or protostellar systems at redshifts z between 10 and 30. Such systems would subsequently mass together into protogalaxies, and these would assemble to produce well-formed galaxies.

Stellar and galactic structure formation will be observed and analysed in the infrared and submillimetre region of the electromagnetic spectrum using the NGST, FIRST and ALMA telescopes. One of the most moving episodes in the history of the Universe will thus be revealed: the emergence of the first starlight and the first material forms.

But darkness will not be absent from this blossoming. It is becoming clear that the formation of supermassive black holes, with masses between a hundred thousand and a hundred million times the mass of the Sun, resulting from the unrestrained collapse of baryonic matter, must influence the genesis and prehistory of stellar societies. The first giant black holes would thus have come into being around $z = 10$. These gigantic structures will very probably be detected in their childhood phase by their effects on surrounding gases. Such effects will be detectable in the X-ray band using highly sensitive instruments like the future telescope XEUS.

By studying the X-ray line of iron at 6.7 keV, we will greatly improve our understanding of the progressive synthesis of metals, and more generally, the chemical evolution of the cosmos. The composition, mass and temperature of the true intergalactic medium, which seems to be dominated by a very hot gas with filamentary structure, should also be revealed by XEUS, a marvellous balcony looking out across the Universe. May we relish the prospect!

Dialectic between stars and clouds

Let us return to the fountain of our own origins, the Milky Way. Astronomy today resounds with discoveries which, although exciting in themselves, only involve tiny details of our island universe. The study of particular objects is of course an interesting exercise, but it would be wrong to lose sight of the overall architecture to which they contribute. Clouds, stars, cosmic rays and magnetic fields are woven together into an intricate tracery that we so elegantly call the Milky Way. It is difficult to put the interior into perspective because the closest objects effectively form a veil and we are faced with a classic situation of not being able to see the wood for the trees.

Buried within its dusty disk (two grains of dust per million cubic metres), our Galaxy is hard to view with any great clarity. Although we are moving along inside it, let us climb out in thought and attempt to draw the picture that an astronomer would see from a hypothetical planet located in one of the galaxies of the Virgo cluster. In fact, our star system would appear as a wheel of fire receding at about 1500 km s^{-1}, brighter and redder at the centre than at the edges. A spiralling wave would be discerned, made up of a twirl of blue stars.

Using a spectrograph to analyse light from the stars and gaseous nebulas at different distances from the centre of the disk, compositional variations would be observed. For example, the oxygen content would be seen to decrease in moving out from the centre towards the edge.

A luminous bulge would be found to adorn the centre of the system, a kind of stellar prominence or swelling. Red globules would be seen to swarm around the flattened disk. This picture would be so easy for the extragalactic astronomer to observe with the appropriate telescope, whilst human beings have had to build it up from scratch, piecing together millions of observations fixed up where necessary with ingenious theoretical extrapolations. Completely immersed within our star system, no global perspective will ever be available to us.

Let us escape from our starry prison. In sentimental mood, astronomy inspires the same kind of emotions as the Valley of the Kings. The magnificent architectonics of star societies are astounding and exhilarating.

Visible light from galaxies similar to our own is mainly emitted by stars, especially the youngest, the most massive and the bluest amongst them, and the gas clouds which they illuminate. Galaxies forming stars at a high rate, such as the Large Magellanic Cloud (LMC), will seem bluer and less uniform than the Milky Way and other spiral galaxies. Those that have fallen to the lowest levels of star formation will appear uniformly reddened.

Naturally, the first category abounds in cold and dusty clouds. We know from having studied such objects at close quarters that these clouds give birth to generation after generation of stars. The second category of galaxy on the other hand stands out by the lack of such fertile nebulosity. Why have some galaxies transformed virtually all their gas content into stars whilst others seem to have conserved their gas over all these billions of years?

It is their shape that gives us a clue. All galaxies probably comprise a flat disk and a spherical halo. They then differ according to which of the two components is pre-eminent. The disk is the place where stars are currently forming, whilst the halo attests to past activity, being a gathering of ancient stars. Only the smaller and hence redder of these ancient stars still shine, because the massive and ephemeral blue stars have by now burnt out.

The disks of elliptical galaxies are more or less invisible, just like the haloes of the irregular galaxies. For its part, the Milky Way sports a prominent disk, crowned by an equally distinguished halo. Bright stars and ionised gases trace out the spiral arms that so beautifully ornament the disk. Old stars, grouped into globular clusters, crown the precious disk.

As already mentioned on several occasions, two types of star can be distinguished by their chemical composition, their motion (velocity) and their membership of the galactic halo or the disk. The first population is old and poor in metals. These are witness to the epoch when the newly born galaxy still sought its final shape.

Galaxies are the structural units of the Universe, the stones used to build up the Cosmos. So, rather than asking how old the Universe is, we might begin by asking how old the galaxies are, starting with our own. Just as a man is as old as his bones, we might say that the Galaxy is as old as its globular clusters, that is, between 12 and 14 billion years.

The total mass of the Galaxy is about 10^{12} $M_\odot$. Of this 1000 billion solar masses, only about one-tenth is actually visible. This is what is implied by the Galaxy's rotation curve, i.e. the graph of its rotation speed at different distances from the centre. All other matter is therefore classed as dark matter. The mass of stars is thus about 100 billion solar masses, and the mass of interstellar material a few billion more.

The stars are sparsely distributed. If they were raindrops, they would be about 100 km apart.

The interstellar medium comprises mainly hydrogen and helium, divided up into clouds of bright or dark gas that glitter elegantly. The glitter is tiny grains of solid matter. The mass of dust is only one-hundredth part the mass of gas, amounting to about 10 million solar masses. However, its effect on the light from stars and on interstellar chemistry is crucial.

Most of the gas in the Galaxy is contained within the disk and in particular in the spiral arms, hence in a layer only a few light-years thick. Although we cannot claim that the space between the stars is empty since the interstellar medium is actually observable, it is not far from being so. It contains on average about 1 atom cm^{-3}, far less than the best laboratory vacuum.

The interstellar medium is thus extremely dilute and highly inhomogeneous. The sparsely scattered material floating between the stars has a mass that only barely exceeds a few percent of the mass incorporated into visible stars. But the interstellar medium bathes in a yet more diffuse entity. Indeed a swarm of fast-moving particles fills it from one side to the other, streaming across in all directions.

This extremely tenuous gas of high-energy particles is superposed upon the other gas of atoms almost at rest which constitute the galactic disk. It includes in its arsenal all the atomic nuclei yet known to us, but also a smattering of elementary antimatter (antiprotons and positrons). These antiparticles are not primordial, but generated by collisions between protons in which part of the proton kinetic energy is transformed into electron–positron or proton–antiproton pairs. What could be more natural? It would, however, be quite astonishing to find even a single anti-helium nucleus. Astrophysics and cosmology would be thoroughly shaken. The AMS (Anti-Matter Spectrometer) is designed to give a clear answer to the question of whether anti-stars exist in the Universe. This project consists in setting up a superconducting magnet aboard the International Space Station to separate matter and antimatter in cosmic rays.

Continually irradiated by photons and bombarded by high-energy particles, space is radioactive. The only relatively sheltered regions are perhaps those at the heart of dense molecular clouds, as attested by their extremely low temperatures (3–30 K).

The gaseous component is made up of a mixture of atoms and molecules which may be either ionised or electrically neutral. Despite its modest proportions, the dust component plays a determining role in the thermodynamics and chemistry of the interstellar medium and in the star formation process that subsequently governs the whole evolution of the Galaxy.

Dust grains act like stones in the desert. They accumulate heat and restore it to the medium in the form of infrared radiation. They are intermediaries between light from stars and interstellar gas, for they absorb stellar photons in a most efficient manner. This is why these clouds appear so dark in photographs. In fact, they shine in the infrared. The dust strewn across the Galaxy trades the big money of stellar light for the small change of the infrared.

The task of the perfect cloud is thus to confiscate visible light by absorbing or scattering it. This collective effect is known as extinction. Such attenuation provides protection against devastating UV radiation. Sheltered at the heart of a dense cloud, molecules can proliferate and put together a whole range of chemistry, called interstellar chemistry. It is often catalysed by the dust grains themselves, for these offer their surface as a reaction arena. Complex organic molecules can thus build up and a thin film of ice may cover the grains.

Interstellar chemistry

Until 1968, astronomers had always assumed that the interstellar medium was essentially made up of atomic hydrogen. Indeed, this ubiquitous element leaves its trace in every quarter in the form of a specific radiation line at wavelength

Table 6.1. *Fractional abundance*
of elements relative to hydrogen

Element	Fractional abundance
O	7×10^{-4}
C	3×10^{-4}
N	1×10^{-4}
Si	3×10^{-5}
Mg	3×10^{-5}
S	2×10^{-5}
Fe	4×10^{-6}

21 cm. Then ammonia (NH_3) was discovered near the galactic centre, followed by water vapour and a whole litany of ever more complex molecules, including ethanol (CH_3CH_2OH). There is far more alcohol in interstellar clouds than in all the existing bottles of armagnac!

The radiotelescopes at IRAM (Institut de Radioastronomie Millimétrique), a French–Spanish–German consortium, can lay claim to a good few discoveries of new cosmic molecules over the past ten years. The IRAM interferometer on the Plateau de Bures in France combines signals gathered by five parabolic antennas. It has an angular resolution of 0.5 arcsec at 1.3 mm.

This wavelength corresponds to a transition of the carbon monoxide (CO) molecule which is found right across the Milky Way and in other galaxies. Radioastronomers consider it to be a good tracer for molecular hydrogen, with which it tends to cohabit and which blossoms into stars. For it should not be forgotten that stars come from the cold. Precise maps obtained by interferometry demonstrate the great chemical wealth of the envelopes surrounding evolved stars. Every day the list of detected molecules grows longer so that it includes several dozen at the time of writing (2000).

We owe much to radioastronomy. It has taught us, for example, that the interstellar medium is the site of complex and varied chemistry, quite different to the chemistry we know and practise on Earth. Indeed conditions in space are very special: low temperatures and densities are often accompanied by the effects of extreme radiation. All chemistry taking place in space depends on the cosmic abundances of the reagents. The commonest elements taking part in the combinatorial art of atoms are listed in Table 6.1, based on the abundance diagram.

The discovery of a new molecule in the sky involves not only a radioastronomy observation, but a good deal of laboratory spectroscopy and quantum chemistry. Measurements of molecular reaction rates have to be made at extremely low temperatures, including the probability of photodissociation. This is the cold analogue of nuclear astrophysics, whose methods were discussed in an earlier chapter.

The extremely varied physical conditions reigning in the interstellar medium, as regards pressure, temperature, density and different types of electromagnetic radiation present, thus beget a complex and unfamiliar chemistry. The ultimate product is the raw molecular material that goes into the building of planets and even life itself. These presolar molecules are found in the cometary and meteoritic matrix, to our great wonder. Without the dust, though, the evolution of the Galaxy would have been very different, for the development of planetary systems would have been excluded. Dust is the missing link between stars and life. Astrochemistry will perhaps clear the way to astrobiology!

The plight of hydrogen

Clouds of gas in the interstellar medium are called gaseous nebulas. These nebulas are regions of the interstellar medium with above-average density. The proportions of elements in the interstellar medium conform to the abundances in the table, that is, 90% hydrogen atoms, 9% helium atoms and less than 1% heavier atoms, where these percentages now refer to relative numbers of atoms rather than relative mass.

The gaseous nebulas are divided up into dark nebulas, reflection nebulas, HII regions (see below), planetary nebulas and supernova remnants.

Dark nebulas are observable by the fact that they conceal the stars behind them. Indeed, the total darkness of certain regions of the sky in which there appear to be no stars suggests that there must be some kind of screen. If there is nothing there, then there must be something there.

Some are spherical and are in fact trapped within their own gravitational influence. These are the Bok globules. They are fertile regions, propitious for star formation. They are sometimes found buried within giant molecular clouds, themselves the building blocks of the Galaxy. Molecular cloud complexes like the one in Orion, for example, are cold and have a very limited lifespan of the order of 10 million years. This is very short indeed compared with the Sun. As their name suggests, they are rich in molecules such as hydrogen and carbon monoxide.

Reflection nebulas are visually much more beautiful. These are clouds of gas and dust that shine by light borrowed from nearby stars, just as the planets shine by the reflected light of the Sun. Starlight is scattered by dust grains floating in the surrounding gas, thus giving away their presence. These nebulas have a blue tinge, for light is more efficiently scattered at these wavelengths. This is why the sky is blue on Earth. It is in fact the air that tints the sky this way, as Leonardo da Vinci pointed out.

To all intents and purposes, galactic space can be assumed to be filled with hydrogen at an average density of 1 atom cm^{-3}, although certain localised regions are much denser, so that others are less dense but more extensive. When the density exceeds a certain critical threshold, hydrogen atoms join up pairwise to form hydrogen molecules (H_2). The densest region of the Orion nebula contains 1 million cm^{-3} hydrogen molecules, or even more.

Localised regions of ionised hydrogen are called HII regions, indicating that the hydrogen atoms are in their second state of ionisation. They appear everywhere in which neutral hydrogen atoms (HI) are exposed to photons with energies above 13.6 eV, because this is the binding energy holding the electron onto the hydrogen nucleus. High-energy photons ionise hydrogen atoms to form protons (H^+) and electrons. HII regions are thus bright, ionised regions around massive, young stars of types O and B. Their spectra are dominated by emission lines.

By studying radio and optical spectra from HII regions and planetary nebulas, to be discussed immediately below, we may establish the abundances of several elements, in particular, helium, absent from the solar spectrum,[1] a point of great cosmological significance, but also nitrogen and oxygen.

Planetary nebulas

Planetary nebulas are so named because some of them, when viewed by telescope, vaguely resemble planets. In reality they are a thousand times bigger than the whole Solar System and bear no relation whatever to planets. The most famous among them is the Ring nebula in the constellation of the Lyre (M57 in the Messier catalogue). Other wonderful examples can be found on the Internet at the European Southern Observatory website (see Appendix 8).

A hot star at the centre is surrounded by a bright shell of gas that has detached itself from the central star. The gas shines by fluorescence. It absorbs ultraviolet light from the small, hot central star (white dwarf) and re-emits it in the form of visible light. The big banknotes (UV photons) are exchanged for smaller

---

[1] The solar atmosphere is too cool to lead to helium excitation, and hence helium lines, except in exceptional conditions associated with flares. The helium content of the Sun is therefore inferred from observations of hotter stars.

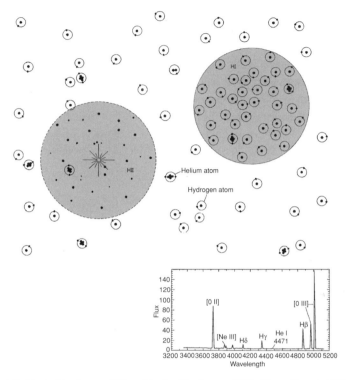

Fig. 6.2. Neutral hydrogen (HI) and ionised hydrogen (HII). (From Pasachoff 1977.)

denominations (visible photons). This picture prefigures the Sun's own death. Our planet will be swept by a hot wind and the atoms of all the dead deposited in the earth will once more belong to the Sun. Born in the stars, the atoms will return to the stars.

These nebulas are similar in some respects to the HII regions. The difference is that here the source of ionisation is an ageing star (white dwarf) in its death throes rather than a strapping young blue star. The fluorescent region is both denser and chemically more complex for it includes those atoms expelled from the envelope of the dying star in the form of a stellar wind.

These superb gaseous corollas thus owe their appearance to the UV radiation emitted by the hot and compact central star. UV photons excite and ionise atoms in the nebula. When the electrons cascade back down to their original energy levels, photons are emitted at wavelengths in the visible range of the electromagnetic spectrum. The blue–green colour of many planetary nebulas is due to emission lines of doubly ionised oxygen at 500.68 and 495.89 nm. These objects have a characteristic temperature of 10 000 K. The expansion speed of the gas, measured through the Doppler effect, is typically between 10 and 30 km s^{-1}. Given that the maximum radius is 0.3 parsecs, their age can be estimated at about 10 000 years. After 50 000 years, all the gas has dissipated into the interstellar medium. The planetary nebula phase is a fleeting moment in the life of a star. About 2000 such objects have been recorded. However, we only see the nearest ones. There are estimated to be about 50 000 across the whole Galaxy. If each one contains on average 0.5 $M_\odot$, planetary nebulas restore several solar masses of wrought matter to the Galaxy as a whole. This matter is enriched in helium and nitrogen, the ash produced in the nuclear combustion of hydrogen.

Supernova remnants and a bright crab

Supernovas are divided into two categories depending on the presence or absence of hydrogen lines in their light spectra. Those of the first type leave no compact residue.

The archetypal Crab nebula results from the second type of supernova, for it encloses a neutron star at its heart, the corpse of an exploded star that can now be admired as a pulsar. Of all supernova remnants, the Crab nebula is the best known, located in the constellation of Taurus. The brightening that followed the explosion was observed by Chinese star-spotters on 4 July 1054, whilst in Europe, the schism was tearing the Church apart. Today, almost a thousand years after the signs of the cataclysm reached Earth (the explosion itself took place 6500 years before that, since the Crab nebula is located 6500 light-years away), the nebula resulting from the explosion is moving out from its centre at a speed of 1500 km s^{-1} and shines as brightly as 80 000 suns. Most of the radiation is emitted in synchrotron form. This particular kind of radiation is produced when high-energy electrons are trapped inside magnetic fields, as we said earlier.

The electrons lose energy by radiating and so have a limited lifetime. Tired electrons must somehow be replaced. The continuous source of relativistic electrons is in fact the central pulsar. Indeed, at the centre of the expanding nebula is enthroned a rapidly spinning neutron star (turning at some 33 revolutions per second), as witnessed by the punctuated message we receive on Earth. This star is clearly an excellent electron accelerator.

A second example of an explosion remnant is the fine lacework of the Cygnus loop, located 2500 light-years from Earth. In supersonic expansion, the gas produces shock waves that excite and ionise interstellar matter, causing it to glow.

The radio wave synchrotron emission is non-thermal in origin. However, a large part of the radiation from supernova remnants is thermal, for heat is generated by the passage of the shock wave from the explosion. These remnants radiate most of their thermal energy in the form of X rays. Emission lines stand out from the spectrum, revealing the presence of magnesium, sulphur, silicon, calcium and iron. Unfortunately, in such agitated conditions, it is difficult to deduce reliable abundances for these elements in the smoking debris of the stellar corpse.

Cosmic rays

Galactic cosmic rays, as they are generally known, are made up of fast-moving electrons and nuclei which plough through space in all directions. At such speeds, the nuclei are stripped of all their electrons and their electrical charge is thus that of the nucleus itself. In particular, these charges are positive, e.g. +6 for carbon and +26 for iron. Being electrically charged, they are deflected by the assortment of magnetic fields that wind themselves in and out across the Galaxy. This means that their final direction when they arrive in our detectors bears no relation to their initial direction, unless they have an extremely high energy. In this sense then the study of cosmic rays is not genuine astronomy.

Galactic cosmic rays are indeed galactic. Together with the magnetic field which confines them, they constitute an essential part of the overall energy content of the Galaxy. Indeed, their combined energy density of 1 eV cm^{-3} is on a par with that of the stars and interstellar gas. These cosmic rays exert a pressure which must be taken into account when analysing the large scale equilibrium of the Galaxy.

The study of galactic cosmic rays is perhaps more an exercise in taste than in visual appreciation. In fact we determine their composition without ever really seeing them. However, they constitute the only sample of matter in our possession that comes from outside the Solar System. The chemical and isotopic composition of this sample is measured using balloon- or satellite-borne particle detectors, since the Earth's atmosphere is fatal to them. When they slam into nuclei in the air, they fragment into tiny particles, thereby losing their original identity.

Composition and energy distribution (energy spectrum) are the two clues that help us, after a long and painstaking enquiry, to understand the sources of these rapid nuclei and determine the mechanisms which first accelerated them.

The lithium, beryllium and boron content of cosmic rays is particularly high. Indeed, the ratio of Li + Be + B to C + N + O is 0.25. We may take this as a direct proof that spallation occurs in space. On their way from source to detector, these fast-moving nuclei are occasionally involved in accidents, that is, collisions with the nuclei of atoms at rest (or relatively so) in the interstellar medium. As a result of such collisions, they may lose one or more nucleons and thereby change their identity. Hence, in a collision with a proton or helium nucleus, a fast carbon-12 nucleus may lose a proton to become boron-11. The ratio of boron to carbon gives an estimate of the quantity of matter encountered along the path.

In order to reconstruct relative abundances of these nuclei at source, we must first expurgate all the fragmentation debris. This is done with the help of a model to be described shortly. The Galaxy is not totally closed as regards cosmic ray movements. Three dangers await any particle launched at high speed in the Galaxy:

- deceleration and subsequent incorporation into the interstellar medium;
- escape into extragalactic space;
- death or rather mutilation in a traffic accident.

The velocity of the nucleus after the accident and of the fragments that accompany it is the same as that of the nucleus before the accident. As we have already mentioned, the exclusion of certain particles from the thermal community, which is essentially democratic even though some particles possess three or four times as much energy as the others, followed by their election to the title of cosmic rays, requires some form of selection and energy enhancement. This selection seems to be made according to atomic criteria.

It is observed that elements with less tightly bound electrons, in other words, those that are more easily ionised, are most strongly represented within the cosmic ray population. This already suggests that the sorting mechanism is essentially electromagnetic. It is thought that only charged particles are brought within the actual acceleration region.

The main accelerating agent, able to raise nuclei to energies that sometimes exceed the performance of our best particle accelerators, is thought to be the shock wave. This is indeed an extremely effective acceleration mechanism, leading in a quite natural way to an energy spectrum of the observed form. The most powerful shock waves are generated by star explosions and supersonic stellar winds. At the end of the day, it is therefore the stars that accelerate cosmic rays to such phenomenal speeds. The violence of the stars is communicated to certain neighbouring atoms which lose their electrons under the effect of the encounter. And so the cosmic rays are born.

Imagine two converging walls filled with holes. A ball is thrown at the middle of one of the walls. By consecutive collisions with the walls, it acquires increasing energy, unless it escapes. For the ever-rarer balls that remain be-tween the walls, the collision frequency increases as the walls move together, and with it the probability of escape. The growing energy also raises the colli-sion frequency. The maximal energy attained depends on the shape and hardness of the walls and on how long the phenomenon lasts. Indeed, beyond a certain energy, the particles are transformed into genuine missiles, capable of pierc-ing the walls. Once the confinement capacity of the walls has been exceeded, the acceleration mechanism reaches an end. Particles escape into space with a statistical energy distribution (spectrum) that falls off extremely rapidly with increasing energy. In fact, the number of particles with energy E is proportional to $1/E^{2.7}$ above 1 GeV. This acceleration mechanism resembles a game of table tennis. The particles are sent back and forth for as long as the round may last. At each impact, their energy increases slightly so that, after a large number of exchanges, they may have acquired a considerable energy.

Protons, nuclei and electrons in cosmic rays would appear to inherit their energy spectrum from this mechanism, associated with shock waves. A non-thermal component in X-ray emissions from the recently discovered remnant of the 1006 supernova provides direct confirmation.

Shock waves abound in the Galaxy. They are produced by any supersonic movement of matter. The average speed of sound in the interstellar medium is around $10 \, \text{km s}^{-1}$ and it is not unusual to find it exceeded by a wide margin. The speed of the stellar wind from massive stars, in particular, Wolf–Rayet stars, can be as much as $2000–3000 \, \text{km s}^{-1}$. Matter ejected from supernovas can attain $10\,000 \, \text{km s}^{-1}$. By the mechanism described above, part of the kinetic energy carried by a whole population of particles is transferred to a few rare individuals which can thus reach considerable speeds. These are the cosmic rays.

The elected particles do not immediately assume the title of cosmic ray. They must first be injected into the acceleration zone at the right speed, and this also appears to happen selectively. The easiest elements to ionise, and these are also the elements that tend to be found in interstellar dust grains, are those that are best injected. This suggests that some previous process of electromagnetic type is able to raise ionised species to energies of a few hundred keV, and that only then are these promoted to high energies by shock waves via the ping-pong mechanism described above.

Secondary effects from cosmic rays

Cosmic rays like nothing better than to share their fabulous energies. They dole it out in various ways, ionising and heating matter encountered along their path,

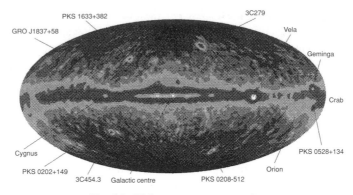

Fig. 6.3. High-energy gamma-ray sky.

i.e. interacting with electrons in the medium. Indeed they tear electrons from nuclei and set them in motion. These electrons in turn deliver some of their new-found energy to the surrounding medium. High-energy protons (1 GeV or more) induce gamma radiation through the intermediary of neutral pions:

$$p + p \longrightarrow \pi^0 \longrightarrow 2\gamma \, .$$

Fast-moving electrons, for their part, spiral through magnetic fields emitting synchrotron radiation, so easily recognisable under the watchful eye of the radiotelescope from the shape of its spectrum. Transferring part of their kinetic energy to stellar or cosmological photons, electrons also produce low-energy gamma photons by the inverse Compton effect. This non-thermal radiation has a characteristic spectrum which makes it easy to identify. Such spectra are found not only in our own Galaxy but also in many others, particularly those with an active nucleus, like the Seyfert galaxies and quasars. We may conclude that the central regions of galaxies are good particle acceleration sites, raising electrons and protons, or more generally leptons and baryons, to high speeds. Electron accelerators seem to be commonplace, whilst the baryon accelerators are rather more scarce.

These then are the actors on the cosmic stage. So let the curtain rise!

The gamma-ray sky

A map of the sky has been drawn up using a spark chamber to record the directions of gamma rays with energy greater than 100 MeV (Fig. 6.3). The result is a strangely unfamiliar sky where a very small number of sources emerge from a predominant Milky Way. A whole population of gamma-ray sources were discovered by the EGRET (Energetic Gamma-Ray Experiment

Telescope) experiment aboard the GRO satellite. This experiment was designed to detect high-energy gamma rays (>30 MeV). EGRET was equipped with a spark chamber in which gamma rays transform into electron–positron pairs. Such pairs are easily detectable because it is always simpler to spot charged particles than neutral particles. The energy and direction of high-energy gamma rays were measured. This gamma radiation arises for the main part from neutral pion (π^0) formation in collisions between protons. Such π^0 particles each decay into two high-energy gamma photons.

The EGRET all-sky map, like that of COS B before it, clearly shows that high-energy cosmic rays are present right across our Galaxy and with comparable intensities in different quarters. This survey of extremely violent reactions shows where in the Galaxy and the Universe the highest-energy collisions take place. A few point sources stand out in the gamma-ray Galaxy. These gamma stars are mostly pulsars, which are nothing other than magnetised neutron stars resulting from massive explosions. One of these gamma stars, discovered by Félix Mirabel at Saclay on the gamma-ray survey drawn up by the SIGMA satellite, is actually a microquasar. These objects are so called because, without being as luminous as their extragalactic counterparts, they do exhibit the same features, namely, jets and apparently superluminal motions. However, a certain fraction of the recently discovered gamma-ray sources remain unexplained, a thorn in the side of high-energy astronomy.[2]

[2] For further discussion, see Paul (1988) and remarks by Isabelle Grenier in *Nature*, April 2000.

7

Histories

Glossary

accretion gravitational attraction and accumulation of matter by a compact object

astronomical unit (a.u.) a unit of length in astronomy, equal to the distance from the Sun to the Earth, or roughly 150 million km

coronene a polycyclic aromatic hydrocarbon (PAH) with chemical formula $C_{24}H_{12}$

gravitational collapse supernova supernova drawing its energy from the core collapse of a massive star

Heisenberg uncertainty principle the principle that when particles are more and more localised in space, the spread of their momenta increases, and vice versa

helium flash nearly explosive inception of helium burning in the dense core of a red giant star

light curve changing luminosity of a celestial object as a function of time

neutronisation transformation of protons into neutrons by electron capture

Pauli exclusion principle the principle that two electrons or two neutrons can only coexist in dissimilar states of motion

planetary nebula misnomer used to describe the gaseous remnant of stars with masses between 1 and 8 solar masses

resonance enhancement of a reaction probability over a narrow energy band

thermal pulses convulsive stage in stellar evolution

thermonuclear supernova thermonuclear explosion of a white dwarf that has been feeding on a companion star

History of the Sun

Nebulous birth

The sky is no empty arena and stars are not the only actors. The other player in the cosmic drama is the cloud.

The business of the perfect interstellar cloud is to confiscate or at least filter the light of stars lying behind or even within it. Certain clouds referred to as bright nebulas are lit up from within. They are in the process of giving birth to a generation of stars, for like rats, cats and fish, stars are born in broods. Hence, the large, dusty and icy interstellar clouds are not only repositories for the ashes of defunct stars, but also for the material that will give body to new stars. Those stars currently forming, still buried deep within this cloudy placenta, can be observed in the radio, millimetre and infrared regions. Indeed, absorption by gas and dust is minimal at these wavelengths.

Still curled up at the heart of the parent cloud, the stellar embryos attract more matter in order to embark upon the visible phase of an object of fixed mass in hydrostatic equilibrium. They then disperse any surrounding matter and begin their own lives as free and independent stars.

In truth, star formation from molecular clouds is no easy subject to study. This is because the processes involved change the density from 10^{-23} g cm^{-3} to about 1 g cm^{-3} within a space of only a few tens of millions of years. Only the force of gravity, whose long range plays a key role, is able to produce such staggering compression rates.

Protostellar evolution and the gradual dissipation of excess gas envelopes are still poorly understood. Here is what may be described as a plausible explanation. The growth of the stellar embryon is driven by accretion. Due to its rotation, a luminous disk forms around it and the system revolves in a cavity some 10 to 100 a.u. across. This cavity is surrounded by huge shells of matter which feed the disk and, through it, the central object itself. They also absorb the greater part of the radiation emanating from inside. These envelopes of gas and dust are dissipated by a bipolar molecular wind, that is, a movement of molecules out along the axis of rotation of the system. The later development of the parent cloud and its future capacity to hatch out new stars are strongly affected by this violent injection of energy. The large number of jets discovered, in combination with the short duration of the ejection phenomenon, suggests that all stars go through a high wind phase at the beginning of their existence.

But how is it that these objects simultaneously accumulate and shed matter? How can a star form by losing mass? It is thought that the solution to this paradox lies in the wind. Material deposited on the star from the encircling disk dangerously increases the speed of rotation. A centrifugal barrier then opposes any further addition of matter. Hence, the stellar growth process can only continue if the star has a way of moderating its rotational speed, even as further material is still accreting. The stellar wind constitutes an opportune braking system, provided that it evacuates only a small mass but a high angular momentum. As we have seen, gas flows also provide a natural mechanism for dissipating the outer envelope. At the end of this process, a properly constituted star of given mass emerges. If the mass is comparable with the mass of the Sun, the object is a T Tauri star. The physics of protostellar winds is thus intimately connected with the central problem of galactic evolution through the distribution of stellar masses (see Appendix 5).

In the beginning, before the Sun became a star, its only source of energy was gravity. It radiated as a result of its own contraction. Matter was stirred up by strong internal currents. At this point our star was totally convective, for it had low temperature and was thus totally opaque to light. These tremendous stirring forces homogenised the material within, so that it had the same composition at all depths. This is why it is assumed today that the chemical composition of the photosphere is just as it was when our star was born. That is, it represents the chemical composition of the protosolar cloud, uncontaminated by nuclear reactions in the core and hence untarnished by any slag from the fires within.

The luminosity then decreased rapidly from 20 to 0.5 $L_\odot$, where $L_\odot$ is the Sun's present luminosity, whilst the surface temperature stabilised at around 4460 K. The Sun looked like an orange. The convective zone was resorbed and covered the star like a blanket. Although just 1% of the mass, it occupies today 30% of the radius!

Luminous life

Phoebus with its golden mane now settled into a long-lasting period of calm. Half of its hydrogen reserves had been consumed. Helium, ash of hydrogen, was accumulating at its centre. The temperature there was roughly 15 million K and the density about 150 g cm^{-3}.

Hydrogen fusion stabilises stars for considerable periods. It generates energy by transforming hydrogen into helium. However, as far as chemical production is concerned, the results are sparse. The accomplishment of the Sun and many stars like it is a modest affair, resulting in a very small addition of helium to the quantity originally generated in the Big Bang. The freshly made helium remains trapped at the end of the main sequence in the central parts of the stars that produced it.

The fact that neutrinos are emitted during this reaction provides an opportunity to observe directly the nuclear reactions taking place in the Sun's core. But its core is like a safe or an urn in which it zealously guards its own ashes.

Nevertheless, at the end of its life, the Sun will release part of its substance, and the helium produced throughout its long career will be poured out into the interstellar medium. The day-star will not always hide the produce of its nuclear alchemy under the mattress.

> The Sun shines.
> We gain
> While he loses.
> For the Sun,
> To shine is to lose energy.
> The Sun clutches his light,
> But cannot stop himself shining.
> He shines just what is allowed
> By his mediocre transparence,
> His quasi-opacity.
> Brightly, he consumes himself,
> Transforming his heart to ash.

The Sun shells out

The bright garment of the great lantern hides its true nature. Now ageing, under a serene countenance, the Sun still carries a blazing inferno in its core. It owes its great age mainly to its gaseous flexibility. In contrast, the sclerotic heart of certain stars condemns them to explosion or to some other separation of core and envelope.

The day-star is an exceptionally stable star, but it is not eternal. Although it manages its nuclear potential with remarkable finesse, calculations show that in a few billion years it will be forced into inner contortions that will drastically alter the structure of its reactor. When the helium content in its core reaches a high level, it will contract in order to postpone a reduction in luminosity. The matter within it will heat up and thermonuclear fusion of nature's second element will commence, generating carbon and oxygen. The resulting increase in luminosity will make the star very dangerous for humankind, if indeed our descendants survive until then and have not taken refuge elsewhere, both of which are highly unlikely.

Ever since the Sun has been the Sun, its size, surface temperature and luminosity have remained practically unchanged. Solar luminosity has increased by about 30% at the very most over the last 4.6 billion years, and feedback mechanisms have endowed the Earth with the priceless gift of liquid water.

However, our models show that the Sun's luminosity will rise gradually until its hydrogen reserves dry up. This can be explained as follows. As hydrogen is transformed into helium, the density remains virtually unaltered, whereas the number of particles per cubic centimetre in the nuclear core decreases. As a consequence, the core contracts slightly and heats up. This increase in density and temperature has the effect of boosting the rate of nuclear reactions, and this in turn has repercussions for the star's brightness.

The Earth's climatic system can be thought of as a well-adjusted thermostat. It is quite clear that it will not be insensitive to changes in the luminosity of our star. Solar luminosity is expected to rise linearly by about 10% over the next billion years, whilst the surface temperature will increase by 1%. Naturally, supernova specialists would be quite unmoved by such a variation. In their field, luminosities rocket to quite unheard-of values in the merest fraction of a second. But to climate specialists back within the confines of the Earth's atmosphere, an increase of 10% in the Sun's brightness is an alarming prospect.

The slightest growth in the Sun's luminosity, let us say just 0.25%, is enough to cause significant modification of the terrestrial climate, although it is difficult to predict the exact extent. This brings us naturally to the great debate over global warming.

According to numerical simulations, temperature changes are not linear. When the Earth heats up, for example, ice melts. This means that more solar light is absorbed and the atmosphere heats up even more. This effect operates until all sea ice has melted, which requires a 4% growth in solar luminance (luminous intensity per projected area).

The role played by water vapour is much less well understood. Indeed, water vapour is an extremely efficient greenhouse gas. As the air temperature rises, the water concentration grows exponentially and the greenhouse effect assumes alarming and irreversible proportions.

If the solar luminosity increases by 2%, the climate model produced by the Goddard Institute for Space Studies indicates a corresponding temperature increase of 4 °C. Climatologists do not usually make predictions over billions of years so this model has not been pursued. However, we may consider that a 10% increase in solar luminosity would lead to a temperature rise of around 12 °C. The result would certainly be catastrophic. Sea level would rise by some 40 cm as the ice caps melted. With a temperature increase of 21 °C, the ice caps would vanish completely and the climate would be changed forever.

By water we were born, and by its disappearance we shall perish. For water maintains one of the most powerful control mechanisms we know. The Earth seen from space is a blue planet scattered with cloud. The cloudy whiteness of the Earth's face is as vital as its aquatic blue. Cloud cover and ice layers are effective regulators in the short term, but the Earth's main thermostat resides in the relation between carbon dioxide and global surface temperature.

Most of the carbon in the world is held in rocks in the form of carbonates. Global warming would increase the rate of evaporation of water. More rain would fall and strong winds would blow, both factors contributing to greater erosion. The rocks would be soaked with water and this would trigger a chain reaction. Calcium released by erosion of rocks would make its way down to the oceans where it would combine with carbonates and shells in such a way that corals would proliferate. The calcium would thus accumulate on the ocean floor, followed by carbon dioxide, for the drowning of the carbonates would be balanced by a draining of this gas from the atmosphere. So in the end, carbon dioxide would be engulfed deep in the ocean where it would be fixed in carbonate-bearing rocks.

Control has now been lost. The powerful regulatory effect of carbon dioxide which controlled global temperatures for 4 billion years has been blown apart. These are the conclusions reached by James Kasting (University of Pennsylvania), who studies the physical factors defining habitable regions around stars. Kasting adds that, according to numerical simulations of the climate, the carbon dioxide concentration drops to 140 ppm (parts per million) after 500 million years, whereas 150 ppm are needed for the survival of the least demanding C3-type plants. Plants are classified according to the type of photosynthesis they use. Those with the highest carbon dioxide requirements, category C3, happen to constitute 90% of the Earth's flora and also represent most of the Earth's food supply.

It may seem that 500 million years should be sufficient for the C4 plants, which require less carbon dioxide, to replace the greedier C3 plants. Surely the survivors would have time to adapt to the reduced harvest of carbon dioxide? However, once it has been triggered, the disappearance of carbon dioxide cannot be stopped. In fact, quite the contrary! It accelerates. After 900 million years, the C4 plants meet the same problems, according to the modern prophets. No more carbon dioxide, no more plants, no more animals. A dismal prospect indeed.

The apocalypse begins to move faster now. Some 1.1 billion years from the present epoch, water begins to reach the stratosphere, an upper region of the atmosphere. At these altitudes, water is dissociated by UV light. The hydrogen released, too light to be held by Earth's gravity, flies off into space. The ocean evaporates and all remaining life forms are obliterated.

Consequently, well before the Sun runs out of hydrogen fuel, life will take its final bow, at least, if we are to believe these numerical simulations. What will happen to the planet which life leaves behind it? Until very recently, astronomers believed that the Sun would lose enough mass through its red giant winds to ensure that the Earth's distended orbit would not be absorbed by the swelling star. The fate of the Earth thus hangs upon the dying breath of its star. In fact, everything depends on the loss of mass, and this in turn depends on the intensity and duration of the solar wind. If it comes early on, the parched and sterile Earth will escape from vaporisation by the dying star, because the Sun's mass loss will weaken the gravitational bond, allowing the planet to move out to a safer orbit somewhere between here and Mars. It will then remain a silent memorial to the life that once flourished on its surface. If the Sun's mass loss occurs later, the Earth will be reduced to ashes and probably swallowed up by the huge red Sun.

As the Sun turns red, humankind will die or fly, but let us not dwell upon this. Let us move on to the very end of our star itself. For in the end the Sun will mutate from a seething red giant to a sleeping white dwarf.

White dwarf

The Sun will eventually lose its gaseous perfection because the stellar core is not immune to sclerosis. Quantum degeneracy is the root of this change in low-mass stars when they have completely burnt the hydrogen reserves at their centre. They are then well on the way to becoming white dwarfs. Let us now analyse this corpse-like stiffness.

For some, the use of the verb 'to expire' to describe the Sun would not invoke the throes of death, but rather a simple stellar sigh. It is true that, when Phoebus leaves the stage with golden mane flowing, this exit will bear no resemblance to the apotheosis of a supernova. Its discreet final transfiguration into crystal and smoke nevertheless conceals a remarkable subtlety. In order to describe how the core collapses and the envelope is discarded, science has called upon its best star modellers. By numerical simulations on their computers, they are able to reconstruct the history of a star of any mass, from its nebulous birth to its shady death, to determine the instant of its final departure, and to describe its *post mortem* state, if indeed the word 'death' still has meaning in the kingdom of the stars. It transpires that the life expectancy of a star, and the final form it assumes at the end of its luminous existence, both depend to a large extent on its mass at birth.

Stiffening of matter

When a gas is compressed, it heats up. When it expands, it cools. Anyone who uses a bicycle pump knows this. However, it is no longer true for electron or neutron gases at very high densities. This deviation from the ideal gas laws has a catastrophic effect on stars. Moreover, the behaviour of a photon gas with regard to volume changes differs from that of a typical gas made up of atoms, ions or molecules. A photon gas does not have the same temperature–energy density relation as a gas of material particles.

From the thermodynamic standpoint, the basic components of stars can be considered as photons, ions and electrons. The material particle gas (fermions) and the photon gas (bosons) react differently under compression and expansion. Put n photons and n material particles into a box. Let R be the size of the box (i.e. a characteristic dimension or scale factor). The relation between temperature and size is $TR = $ constant for the photons and $TR^2 = $ constant for the particles. This difference of behaviour is very important in the Big Bang theory, for these equations show quite unmistakably that matter cools more quickly than radiation under the effects of expansion. Hence, a universe whose energy density is dominated by radiation cannot remain this way for long, in fact, no longer than 1 million years.

Returning to the stars, the problem is to decide which of the photons, ions or electrons are going to dominate and thus determine the pressure. In fact, the determining role falls to the photons at high temperature and low density, to the electrons at low temperature and high density, and to the ions in the intermediate situation.

A gas in which the pressure no longer depends on the temperature is said to be degenerate, an unfortunate term indeed, because the corresponding state borders on perfection. One might call it a state of perfect fullness, since no interstice is left vacant. Electrons occupy all possible energy states and total order prevails. Both the electrical conductivity and the fluidity also attain perfection. Objects made from this sublime form of matter are perfectly spherical. And yet, in quantum circles, this state of nature is obstinately referred to as 'degenerate'!

Hence, above a certain density, stellar matter manifests quite different properties which can only be described by quantum mechanics. Electrons in the medium begin to oppose gravity in a big way through their exaggerated individualism. In fact, elementary particles with half-integral spin, such as electrons, neutrons and protons, all obey the Pauli exclusion principle. This stipulates that a system cannot contain two elements presenting exactly the same set of quantum characteristics. It follows that two electrons with parallel spins cannot have the same velocity.

To explain this behaviour, physicists appeal to the very foundations of quantum theory. Because of their much reduced freedom to move in space, the particles can be considered to be more and more localised. Then, by Heisenberg's uncertainty principle, the spread in their velocities has to grow. In other words, some particles may have much higher velocities than those allowed by the temperature. A quantum pressure arises at high densities, when the mean distance between electrons becomes comparable with their associated wavelength

$\lambda = h/mv$, where h is Planck's constant, m the particle mass and v the particle speed. As the speed is proportional to the square root of the temperature ($mv^2/2 = kT$), we see that the quantum effect is much more pronounced at high densities and low temperatures, and when the particle in question is very light. The pressure then becomes independent of the temperature. Conversely, for a given density, the quantum effects disappear above a certain critical temperature and the stellar material reassumes its initial flexibility.

In this physical state, the onset of nuclear reactions can have explosive consequences. It is believed that nuclear combustion in a degenerate medium is responsible for the so-called helium flash that shakes small, ageing stars, but also for type Ia supernovas (to be discussed shortly). There are two types of star in which quantum pressure counterbalances the force of gravity, viz. white dwarfs and neutron stars. In the first case, the pressure is exerted by electrons, in the second, by neutrons.

Star, tell me how much you weigh and I will tell you how long you will shine and how you will make the transition from luminous to dark.

The Sun is calm today, because each chunk of its vast body is simultaneously attracted towards the centre by gravity and repelled towards the outside by heat pressure. The solar nuclear power station is self-regulating. However, the nuclear fire, operating through thermal pressure, cannot oppose collapse forever. When the fuel is exhausted, the fire will go out and relentless gravity with take possession of its ashes.

All stellar evolution can be summed up by a simple rule: the star tries to make itself as small as possible. Its life story is one of contraction, but in a discontinuous manner, with sometimes long pauses during which it maintains its size. There are phases when the outer layers are driven off by radiation pressure (stellar winds, ejection of the envelope) and brief periods when the star violently readjusts itself, but without breaking apart (helium flash, thermal pulses).

Its past, present and future countenance can be calculated using star models, and the results can be compared with observable stars. This celestial object corresponds to a young Sun, and this to a Sun in the throes of death. This so-called 'planetary' nebula, so beautiful to look at, is nothing other than a Sun torn apart.

For a whole range of stellar masses between 1 and 100 times the mass of the Sun, evolutionary tracks are traced out in the temperature–luminosity plane, also known as the Hertzsprung–Russell diagram, so frequently referred to by astronomers.

As time goes by, an imaginary star with the mass of the Sun follows a well-defined path which distinguishes it from all others by its shape and the speed with which it is travelled. The point representing our own Sun is located on what astronomers like to call the main sequence, a diagonal bar across the temperature–luminosity plane. Like most visible stars, it is transmuting hydrogen to helium in its core. In the coming 2 billion years, a 40% increase in its luminosity will cause a serious greenhouse effect on our planet. The warning signs were described above. There will no longer be any winters. The trees will burn, the oceans will be brought to boiling point, and the atmosphere will grow opaque. The Earth will become like Venus. In 3.5 billion years, when all the hydrogen has been consumed in the deepest confines of the day-star, it will leave the peaceful main sequence and pass into the turbulence of the red giant branch. Life will long have disappeared from the surface of the Earth, but humankind, I guarantee it, will have mastered interstellar spaceflight.

Now that there is no more hydrogen at the centre of the star, its evolution must necessarily speed up. The central regions contract quickly, for nuclear power production has ceased and gravity is no longer counterbalanced by thermal pressure. But around the region where the hydrogen has run out, nuclear reactions continue, and ash adds to ash. The mass of helium grows and grows, and the combustion zone moves out towards the surface. The helium core, deprived of the heat it needs to hold itself up, contracts still further, releasing even more gravitational energy. The star as a whole must modify its structure to evacuate the surplus energy. It improves energy transfer by means of convection, and cyclopean currents develop. The outer convective zone extends rapidly towards the centre of the star to take control of three-quarters of its mass. But convection cannot transfer all the energy released, so the outer layers expand to let through still more. As the star's radius increases, its temperature drops and the photosphere turns red. The point representing the star on the HR diagram slips to the right to join the red giants.

The Sun reaches a luminosity 2300 times greater than its current value and a radius 170 times greater. The planet Mercury is swallowed up. The solar wind is now much amplified and our star loses 38% of its mass into space.

After pausing for 110 million years on the officially termed horizontal branch (a mere point on the scale), right through the phase of core helium burning, our star will suddenly climb up the asymptotic giant branch. Four thermal pulses will then occur. In the first, it will reach its maximal size of 213 $R_\odot$, or 0.99 a.u. However, the Earth and Venus will not be engulfed, for the Sun's mass will by then have fallen to 0.60 $M_\odot$ due to evaporation effects. The two planets will thus profit from a weakening of their gravitational bonds, gravitating out to distances of 1.2 and 1.7 a.u., respectively. The face of the Earth will be drastically different. The temperature will soar to 2000 K and the mountains will melt. The solar luminosity will culminate at 5200 $L_\odot$ during the fourth thermal pulse.

Throughout the final phases, the solar wind will sweep across the Earth, tearing up the surface. The atoms of our dead will be driven from below and returned to the Sun. The Earth will surrender her dead. Finally, lit up by the bright white glare of the central star, the ejected matter will blossom into a beautiful planetary nebula rather like the one that so delightfully ornaments the constellation of the Lyre. The atoms of all human beings will shine in the sky, mixed with those of the animals and the stones.

Let us revisit the helium core, for this is the heart of the matter. As a result of the relentless contraction, the density continues to increase, and with it the temperature, but at a slower rate now. When the central temperature reaches 100 million K, the density is 10 000 times greater than the density of water. Subject to an imperious quantum principle that forbids them any freedom to overlap one another, the electrons make a final stand against compression and confusion. For this reason, they assume an increasing contribution to the pressure.

Their velocity and through this their pressure are determined solely by the density of the medium. Temperature is no longer relevant. In the heart of the red giant, where quantum pressure exercises sovereignty, gaseous flexibility is lost, and so too is moderation of reaction. Meanwhile, contraction releases large quantities of gravitational energy that must be evacuated. When our star reaches the high point of the giant branch, its core crosses the fateful value of 100 million K. Helium fusion begins. Nuclear energy poured out by the transmutation of helium into carbon and oxygen induces a significant rise in temperature. But the core has lost its former flexibility and hence also its ability to readjust. It can no longer refresh itself by swelling up, and the temperature is free to rise still further. The nuclear reactions race ahead, freeing enormous quantities of energy in a very brief space of time. And yet this profusion of energy leaves almost no trace at the surface. Indeed, the energy produced filters very slowly through towards the distant photosphere of the red giant, and by the time it finally arrives, it has been diluted and softened.

The core now cools very suddenly, as quickly as it heated up, for it rediscovers its gaseous privileges under the effect of this intense heat. In fact, above a certain temperature threshold, thermal pressure moves back into the dominating position. So pressure is once more open to the influence of temperature and control is re-established. Expansion of the core reduces the density by a factor of 50 and the star's evolution recovers its stately pace. Having survived the helium flash, the star begins the second calm and lasting period of its existence. Down in its depths, it placidly transforms helium into carbon and oxygen. Hydrogen is still being transmuted in a spherical shell contiguous with the helium burning core.

After 12.3 billion years of solar existence, all the helium at the centre has been consumed. The production of nuclear energy comes to an end and a further contraction is induced. The density rises to the point where the quantum pressure of the electrons becomes dominant again. Out on the edge, hydrogen continues to burn in a thin shell. Then helium starts to burn in an inner layer and, as before, the outer regions of the star react to this upsurge of energy by increasing their volume. The star is once again a red giant. It now contains two nuclear fusion zones in the form of concentric shells. The inner one is home to helium burning, the outer one to hydrogen burning. This situation is not altogether restful. Sudden flares called thermal pulses suddenly raise the star's luminosity. The Sun is then located on the vertical region of the HR diagram known as the asymptotic giant branch. Under intense pressure from radiation, the envelope is forced away in gasps. The bloodless star deflates to just 3% of the Sun's current radius. In 50 000 years, it crosses the whole HR diagram, its surface temperature changing from 4000 to 10 000 K and its colour from red to white. Nuclear reactions fail in the disburdened and callous remnant. The little white star may now illuminate the ejected gases, portraying them with shimmering colours.

In 7.7 billion years from now, a white dwarf will have taken the place of our star. A solid and compact object no larger than the Earth, this star will be maintained by the stubborn pressure of electrons struggling to preserve their living space, according to a well-established quantum tradition. The tiny crystalline Sun, slowly cooling off, will keep on radiating for billions of years before finally fading into darkness. The extinguished Solar System, deprived of Mercury and perhaps the Earth, will count as dark matter. Half the atoms of the former Sun will have been crystallised into the final ember and thereby removed from the future evolution of the Galaxy. But the other half will wander off into space. The atoms of the Earth's surface will participate in this diaspora. Atoms carried away by the fiery winds of the Sun will be incorporated into new stars, new planets and maybe new human beings.

Let us review all this. At an age of 12 billion years, the yellow dwarf will mutate into a red giant with 100 times the radius. This in turn will give way to a vaporous cloud with a very small white object at its centre. When the yellow dwarf has handed over to the white dwarf, the solar pageant will be close to an end. But for the moment, the Sun still shines staunchly.

Shining relentlessly

Once again we are astounded by the stars' longevity and perseverance. It would seem that they like to shine for a long time and so pursue their glowing career for as long as possible, postponing extinction or disintegration until the very last. What are these beings that shine so obstinately? Indeed, are they beings? Stars are evolving and reproducible entities, but this does not mean that they are living. It is an abuse of language, or at best a metaphor, to say that stars are born, live and die. Rather, they form, evolve, then cease to shine. However, the anthropomorphic vocabulary fits the star so admirably that it would be a loss to language if we were to banish it from the realm of astronomy. I will therefore abandon myself here and there to a little stellar prosopopoeia, and not, I may say, without taking a secret pleasure.

The impressive longevity of stars is in no way due to some intention on their part. It arises from a series of physical causes in combination with a remarkable structural flexibility. The life work of the perfect star consists in shining, but not too much. For let us repeat: the energy of each photon emitted must be replaced. A perfectly transparent star would instantaneously empty itself of all energy. It would burn like a hay bale. The weak interaction (radioactivity), which would be better referred to as the slow interaction, is also a factor contributing to the star's longevity, for it considerably slows down the thermonuclear fusion of hydrogen through its hold over the initial reaction

$$\text{proton} + \text{proton} \longrightarrow \text{deuterium} + \text{positron} + \text{neutrino}.$$

Without it, there would be no neutrons and hence no atomic nuclei. Furthermore, it is responsible for neutrino emissions, for it is the weak force which governs the transformation of protons into neutrons, and this can only be beneficial for nuclear stability.

So there is opacity, that is, the fact that stars retain their light in such a way as not to instantaneously empty themselves of their energy. But apart from this, another determining factor behind stellar longevity is the extreme sluggishness of the two reactions that initiate hydrogen and helium fusion. The reasons in the two cases are very different. The first cause of slowness arises from the very nature of the weak interaction, which is required to transform a proton into a neutron during a fleeting collision between two protons if the reaction product is to be viable. The second relates to the rarity of three-body collisons needed to transform helium into carbon, although it is so easy to say that three times two equals six.

History of carbon

Stars and life

The space between stars is not empty, but filled with matter composed mainly of dust and gas gathered up into colossal clouds. In our own Galaxy, these clouds are spread across a thin layer, narrow but extremely extensive. Seen from above, it takes the form of an elegant spiral. Gas, dust and radiation are the three main components of this Galaxy, and of all galaxies.

The first faltering steps of molecular astronomy were intimately related to the birth of modern spectroscopy. It was the discovery at the beginning of the nineteenth century that the Sun and stars are composed of the same elements as the Earth, which led astronomers to the idea that spectroscopic techniques can be used to observe cosmic chemical processes.

The close agreement between laboratory spectra and observed spectra is an unshakeable proof of correct identification. Molecules are mainly detected in the submillimetre range and the adjacent infrared fringe of the electromagnetic spectrum.

A hundred or so different molecular species have been spotted in molecular or circumstellar clouds. Most interstellar molecules are organic molecules, that is, they are carbon-based. This indicates that chemical evolution does occur on a cosmic scale. What is more, many molecules on the interstellar list are fundamental building-blocks for the construction of biological structures.

Molecules observed in the gaseous phase and on the surface of cosmic dust grains are very probably produced from smaller atoms, ions or molecules by local chemical processes. This is no longer the nuclear chemistry we have been discussing up to now, but ordinary atomic chemistry which proceeds via electrons and ions. Although organic molecules dominate, there are also free radicals and structures containing functional groups like NH, NH_2 and $COOH$.

The trio carbon, nitrogen and oxygen (C, N and O) are of paramount importance in biological structures. Atomic carbon plays a very special role insofar as its electronic configuration, and in particular its tetravalence (the four outermost electrons being available to form as many bonds) endow it with the unique property of forming long chains and complex ring structures, both essential to life. These properties are well established on Earth, but they also operate in the cold and dilute environment of galactic space.

Life appears at the end of a long chain of events, a long sequence of small steps, each one linked to the next. In the cosmic reference frame, the first four steps are important, if not essential, for the development of life.

1. The beginning of nuclear evolution, launched in the Big Bang, the creation of matter, the emergence of nucleons and the construction of the first nuclei hydrogen, deuterium, helium and lithium that took place in its immediate wake, followed by formation of the first stars in the early Universe and the establishment of stellar nucleosynthesis leading to production of carbon and all the other elements.

2. Molecular evolution which occurs preferentially in a very cold, dilute medium, permanently irradiated by diehard cosmic rays and photons. A tremendous variety of relatively complex molecules has already been produced, but it seems to have stopped well before the creation of interstellar life.

The biggest interstellar molecules found in the gaseous phase contain 11 atoms. Whether there are more developed molecules remains unknown. In any case, there appears to be nothing comparable with the molecules of life.

Great flat, ring-shaped molecules of a special class have been put forward to explain certain incongruous spectral features in the infrared. These are the polyaromatic hydrocarbons (PAHs), built by joining together benzene rings. Examples are naphthalene ($C_{10}H_8$) and coronene ($C_{24}H_{12}$), and there are many others looking somewhat like graphite dust.

One task awaiting molecular astronomy is to find the missing links that separate 11-atom molecules from PAHs and fullerenes with 20 to 60 atoms, whilst a whole world lies between the most evolved molecules and a simple dust grain which contains, at the lowest estimate, 1 billion atoms.

The key problem for physicists studying the interstellar medium is to establish the maximum length of linear carbon chains and then to find out from which length these chains tend to close up and assemble into plane rings and three-dimensional fullerenes by spontaneous polymerisation.

3. Prebiological evolution which gives rise to macromolecules capable of organising hereditary transmission, thereby bridging a further gulf between inanimate and living matter. This stage of evolution must synthesise complex molecules such as amino acids, polypeptides, proteins and DNA that could make up the genetic alphabet. Astrobiology has not yet come into being. Although science has made tremendous progress in this field, there is a long way to go.

Today it is impossible to say whether prebiological evolution may have occurred in space, but for the moment there is no reason to exclude this option. Whatever hypotheses are retained, they must surely focus on the warmer regions (if we may use this word!) of the interstellar clouds, the very regions where star formation occurs. However, intense astrophysical searches for the simplest amino acid, glycine (NH_2CH_2COOH), in the dense cores of interstellar clouds have so far failed.

4. Biological evolution itself seems to be associated with a much warmer, denser and damper environment than can be encountered in these clouds, corresponding rather to planetary conditions. In such surroundings, the synthesis of macromolecules can proceed at a much more lively rate. These

are responsible for maintaining cellular structures since they convey genetic information. Proteins for their part are built up from amino acids. The relation between the DNA sequence and the corresponding protein sequence is called the genetic code. Establishing this code is a fundamental step towards information transfer and hence the formation of life.

But do we know what cunning enterprise the sky has been engaged upon in order to prepare and mould this bounteous carbon and then to extract it from the great stellar furnace? We shall once again follow through the main avenues of nuclear astrophysics, but this time considering life as we know it, or rather, one of the conditions that makes it feasible: carbon.

A complex network of evolutionary processes has been woven over the 10 billion or so years that have elapsed since the Big Bang. These phenomena, including the emergence of galaxies, stars and planetary objects, but also the preceding build-up of their constitutive chemical elements, can be taken as conditions for the very existence of intelligent life on our own planet, based as it is on carbon. By definition, before carbon had formed, there was not the slightest chance of organic molecules appearing in the Universe. Likewise, the existence of molecules of genuinely biological complexity depends on the presence of other elements such as nitrogen, oxygen, phosphorus and iron. Moreover, a constant source of energy is needed to maintain living beings in existence. This energy is supplied by nuclear reactions which synthesise helium in the Sun. Matter and energy are *sine qua non* conditions for life. Matter comes from massive stars and exploding stars, or supernovas. Energy comes from the Sun, a modest representative of the silent majority of stars.

We have taken the elements that compose us from the sky. Today it is quite obvious. From hydrogen to uranium, the sources of atomic nuclei have been painstakingly established. Let us just say it again: these sources are the Big Bang for hydrogen, helium and a pinch of lithium-7, stars for all the elements between carbon and uranium, and cosmic rays for lithium, beryllium and boron. Just where order and the violence of a mechanical death seemed to reign in a clockwork Universe, we find the life-giving exuberance of the primordial explosion, of stars, and of wandering particles that dance through magnetic fields.

Red giants: from helium to carbon

In an ageing star which has left the main sequence, two regions can be distinguished. The first is a core, largely composed of helium produced by hydrogen fusion. The second is an envelope comprising mainly intact hydrogen. Hydrogen fusion continues on the boundary between the core and this outer envelope.

At low temperatures (15 million K), reactions between helium nuclei are inhibited by electrical repulsion. On the other hand, the nuclear properties of lithium, beryllium and boron nuclei ($Z = 3, 4, 5$), and in particular their stability, are such that they are extremely fragile, decaying at temperatures of only 1 million K. For this reason, they are not formed in appreciable quantities in stars and cannot serve to bridge the gap between helium and carbon, species noted for their nuclear stability but which, it should be recalled, occur only in minute amounts in nature.

If the star is massive enough, the force of gravity resumes core contraction, and this in turn leads to ever higher temperatures and densities. Core contraction is accompanied by expansion of the envelope, for reasons which have not yet been fully elucidated, initiating a new visual stage in the star's life. Its countenance is transformed as it mutates into a red giant.

If a star is not massive enough for helium fusion to ignite, it simply exhausts its hydrogen reserves and stops evolving. As a white dwarf, it then joins the rest of its kind in the great stellar cemetery.

In the red giant phase, the gravitational force ruthlessly pursues its task of crushing the core. The temperature and density increase in symphony, for compressed gases automatically heat up. When the temperature reaches about 100 million K, corresponding to a density of 100 000 g cm^{-3}, a new type of nuclear reaction begins to occur in the centre of stars.

Among all conceivable reactions leading from helium to carbon at such temperatures, laboratory studies and theoretical considerations suggest that only one is in any way probable. Certainly, helium fusion has not been established without difficulty. This reaction takes place in two stages, represented by the equations

$$^{4}\text{He} + {}^{4}\text{He} \longrightarrow {}^{8}\text{Be}^* + \gamma,$$
$$^{8}\text{Be}^* + {}^{4}\text{He} \longrightarrow {}^{12}\text{C} + \gamma.$$

The asterisk on the ^{8}Be indicates the extremely short-lived nature of this nuclear species. Its lifetime is estimated at 10^{-16} seconds. In order for a carbon nucleus to form, three ^{4}He must therefore collide almost simultaneously, hence the term 'triple α process'.

The chances of such three-body collisions are very slight because the mediating ^{8}Be is so ephemeral. This guarantees the red giant phase an enviable longevity of several million years; even this collision probability is amplified by a perfectly chance resonance, to which we shall return shortly.

Stellar nucleosynthesis thereby leap-frogs over three fragile elements, lithium, beryllium and boron, moving more or less directly from helium to carbon. At this stage in the star's development, the elements needed to form biological compounds have already been wrought in the furnace. Carbon is the first of the light elements to be produced exclusively within stars. A whole procession of other elements will follow in its wake.

When helium fusion begins, the core of the star is stabilised and a new spherical equilibrium is set up. Gravitational contraction is balanced by the expansive pressure of heat levels maintained by nuclear fusion reactions. Oxygen is produced to the detriment of carbon via the reaction

$$^4\text{He} + {}^{12}\text{C} \longrightarrow {}^{16}\text{O}^* + \gamma.$$

However, the fresh produce from this bout of nucleosynthesis remains locked up in the depths of the star, rather as in a larder.

To sum up, the star begins by burning hydrogen on the main sequence, converting it into helium. The concentration of helium thus gradually increases. The star's core heats up until the triple α process can form carbon-12. From there, oxygen-16 can also form. At this stage, if the star is at least eight times more massive than the Sun, more complex nuclear reactions can occur. Otherwise, for less massive stars, nuclear activity reaches an end and the star freezes into a white dwarf, rich in carbon and oxygen, after ejecting a large part of its outer layers.

Nuclear finery

In the glory of its Big Bang, why did the Universe not generate carbon in a brief instant? Why did creation not go to completion in the third minute?

The first step towards an answer is once again the fragility and instability of helium's fusion products. Why then is helium so stable whilst its offspring are so fragile? Why are nuclei with masses 5 and 8 times the proton mass so unsure of themselves that they have disappeared from the map of the world? The explanation for this can be found in the microarchitecture of the atomic nucleus, a subject that is hardly conducive to literature.

The second part of the answer is that, due to its expansion, the temperature and density of the Universe fell too rapidly for the junction between helium and carbon to be established via the triple α reaction: 3 helium-4 $\rightarrow$ carbon-12. This leads naturally to another question: why was the Universe expanding so fast? We can only answer that this is how the world is.

The Universe had therefore to invent the star in order to continue its nuclear edifice. Let us return now to the strange story of carbon, gold and life, born it would seem of exquisite nuclear coincidences occurring within a stellar context. The coincidences in question involve the quantum properties of carbon nuclei and constitute conditions for the possibility of human life itself.

A slight detour into the field of nuclear physics will be necessary to prepare the ground. Depending on the circumstances, any composite system adopts different states and configurations characterised by its various energy levels. Nuclei do not escape this rule. The existence of excited states is characteristic of the composite state of the system. A system like carbon consists of a finite number of elementary constituents, namely, six neutrons and six protons, and it develops a finite number of distinct configurations. Internal activity taking place in such structured objects usually leads to rearrangements. This is why we observe in carbon, as in other nuclei, a great many energy levels, like so many different ways of life. A different energy level, measured in keV or MeV, is associated with each configuration, and this proves without the shadow of a doubt that other objects are hiding within the entity 'carbon'. Of course, these objects are protons and neutrons. Each proton and neutron acts upon the others and is exposed to their actions.

So how does nature manage to stick together three helium nuclei? This was the question on every nuclear astrophysicist's lips back in the heroic age. Indeed, two α particles which collide with sufficient kinetic energy to overcome their mutual electrical repulsion form a transitory eight-nucleon structure, beryllium-8. But in the case of this nucleus, it is the repulsive electrical force that wins over the cohesive nuclear force. For this reason, beryllium-8 is spectacularly unstable, quite the opposite of its progenitor nucleus, helium-4, which is a paragon of cohesive virtue. Beryllium-8 requires a mere 10^{-17} seconds to split in two. A third helium nucleus must therefore be adjoined more or less instantaneously if carbon is to come out of the encounter. A simple calculation shows that this process is highly unlikely. The production of carbon, and with it oxygen and all heavier elements, thus remained a mystery until in 1952, Edwin Salpeter, who directed Hubert Reeves's doctoral thesis, suggested that carbon-12 could be produced in two very rapid steps.

Even then, the way was blocked by the fact that the third accomplice as often as not smashed the feeble beryllium-8 in two. Meanwhile, British astrophysicist Fred Hoyle struck upon a quite brilliant idea. To get around the difficulty, he hypothesised that the energy levels of beryllium, helium and carbon had to have very precisely adjusted values in order to accelerate Salpeter's two-step process and give it every chance of coming to fruition. Everything thus depended upon the consonance between energy levels.

Nature's helping hand in this matter came from the fact that the energy levels of the three crucial nuclei could be fine-tuned to boost the reaction probability, with the fortunate consequence that carbon and the heavier nuclei could then be produced in significant quantities. Since energy levels are purely quantum features of nucleon systems, one could only rejoice at nature's delicate touch.

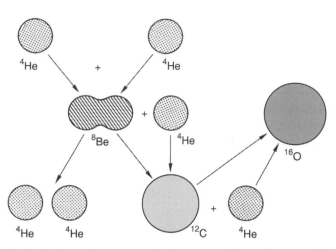

Fig. 7.1. Synthesis and destruction of carbon in stars. Two helium nuclei join to form ^{8}Be which may immediately decay or join with another helium nucleus to form carbon. The latter is then transformed into oxygen.

In order to have a better appreciation of the finer details, let us examine them more closely (Fig. 7.1). When two nuclei collide, the new nucleus serves as a depository for the kinetic energy of the two particles that produced it, together with their mass energy, reduced by its own binding energy. The new nucleus naturally seeks to occupy one of the steps of its own energy scale, that is, a well-defined energy level. If the combined energy of the incident particles (mass energy plus kinetic energy) is not exactly right, all excess energy is used to eject particles from the new nucleus, or if the excess is significant, to break the latter into several pieces. This reduces the chances of finding two nuclei aggregating when they enter into contact, and all the more so for three nuclei. In most cases, the protagonists bounce off each other.

However, if everything holds together, the new nucleus will be created with exactly the right amount of energy, corresponding to one of its natural energy levels. It will then dispose of its excess energy. It will decay by emitting one or more gamma photons as it descends the energy stairway. In this case, the reaction will go through extremely efficiently. This correspondence between the energy of the initial particles and the appropriate level of the residual nucleus, known as a resonance, depends critically on the internal structure of the species involved in the collision and consequently on the fundamental interactions that fashion that structure, the weak, strong and electromagnetic forces.

Hoyle thus understood that the only way to manufacture a suitable amount of carbon in stars was to go through a resonance involving the three helium-4 nuclei, beryllium-8 and carbon-12. The mass energies of the three nuclear edifices are fixed and cannot be altered. The kinetic energies of the three reactants, on the other hand, depend on the stellar temperature and can be calculated. On the basis of this temperature, deduced from models of red giants, Hoyle predicted the existence of an energy level of carbon-12 that had not yet been detected, in harmony with the combined mass energies of beryllium-8 and helium-4. Although skeptical to begin with, Ward Walling of the California Institute of Technology discovered the excited level at 7654 keV above the ground state with the particle accelerator at the Kellogg Laboratory there. This is just 4% above the combined mass energy of beryllium-8 and helium-4. Therefore, for the consonance to occur, the two nuclei had to undergo a relatively gentle collision so that their kinetic energy would be about 4% of their mass energy, that is, some 287 keV. Now the centre of a red giant star provides an ideal site for this so-called triple α reaction for, according to estimates, the temperature must be about 200 million K and the density some 100 kg cm^{-3}. We should note in passing that stellar models were still in a very preliminary stage of their existence. A lot of trust was required in order to present such a firm suggestion. Fred Hoyle's prediction was checked and confirmed by subsequent experiments. What he was asserting was that, since we exist, carbon had to have an energy level at 7.6 MeV above the ground state (Fig. 7.2). This striking piece of reasoning was later to fuel the debate over the so-called anthropic principle, but it is not our intention to discuss that here.

The remarkable nature of the argument should not be underestimated, however. Suppose, for example, that the key energy level had been 4% lower than the mass energy of the ^{8}Be $+ ^4$He system. In this case, the resonance would not have been produced for the simple reason that, although one can always add kinetic energy to the overall balance, one cannot subtract it. We should also mention that, when a carbon-12 and a helium-4 nucleus meet in the core of a red giant, where conditions are right for helium fusion, an oxygen nucleus is very likely to form. It turns out that the sum of the mass energies of carbon and helium is just 1% above an energy level of oxygen-16. But this 1% difference is not enough for all the carbon to disappear in the stellar crucible, thereby destroying any chance of life at a later date.

It would thus seem that we owe our existence to at least two coincidences, one negative and the other positive, involving the energy levels of carbon and oxygen. But this is not all. In order for carbon and oxygen to be scattered to the four corners of the Galaxy, certain stars had to explode. In the same manner, the dissemination of the atoms of life through the cosmos can be imputed to a special adjustment of forces, particle properties and the constants of nature. Apparently our existence depends on a series of coincidences. We know which path we have come along, but we still do not know why.

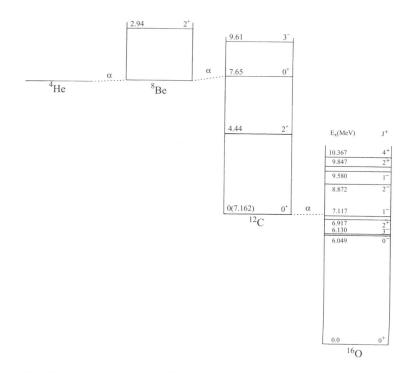

Fig. 7.2. Energy levels involved in helium fusion. The existence of an energy level of the carbon nucleus at 7.65 MeV above the ground state is particularly welcome. It considerably increases the probability of carbon synthesis in red giants.

History of iron

Last dash before oblivion

The destiny of stellar species invites the following question: what mutation transforms an otherwise peaceful and long-lasting celestial body into an explosive and creative supernova? In fact it is the star's luminosity and the ruinous energy expenditure of its glorious years that bring about its own downfall. Indeed, to shine it must burn, and to burn it must perish. In a struggle to draw out their final lustre, massive stars push their fires to the very limit and exhaust themselves in an attempt to preserve their former brilliance.

It is widely believed that when a star moves through the red giant phase, its internal architecture loses the original simplicity it had when hydrogen was burning in its core. It has now become a strange fruit, with its kernel of oxygen and carbon, decked out with shells of helium and hydrogen. The high electrical charges of the carbon nucleus ($+6$) and oxygen nucleus ($+8$) inhibit core nuclear reactions. Deprived of support from heat production in the centre, the core has to contract, raising the temperature even further. When it goes above 600 million K, a new type of nuclear combustion appears on the stage. The first of these are carbon and oxygen fusion, written symbolically as

$$^{12}C + ^{12}C \longrightarrow ^{20}Ne + \alpha,$$
$$^{16}O + ^{16}O \longrightarrow ^{28}Si + \alpha.$$

Insofar as these reactions occur relatively rapidly at the high temperatures now prevailing, the star's evolution accelerates enormously. This is exacerbated by the fact that it suffers a significant energy loss due to thermal neutrino production via the reaction

photon + photon $\longrightarrow$ electron + electron $\longrightarrow$ neutrino + antineutrino,

from the moment carbon begins to burn.

Since combustion of carbon, oxygen and the following elements produces nuclei with masses ever closer to 56, and it is here that the nuclear binding energy culminates, the energy yield has to diminish. Less and less energy is generated per gram of matter burnt. Silicon is less nourishing than oxygen, oxygen less than helium, and helium less than hydrogen.

As a consequence, the time-scale of each fusion cycle is shorter than its predecessor. For a star of mass 20 $M_\odot$, for example, the main sequence (core hydrogen burning) occupies roughly 10 million years, whilst core helium burning lasts only 1 million years, carbon burning 300 years, and oxygen burning a mere 200 days. The star is a red onion whose innermost region is made up of silicon nuclei (Fig. 7.3). At temperatures close to 1 billion K and densities of the order of $1\,000\,000$ g cm^{-3}, silicon combustion gets under way, coming to a close only two days later!

This latter step proceeds via partial photodisintegration. Because silicon has such a high electrical charge ($+14$), direct fusion of two silicon nuclei is extremely difficult, if not impossible. Nucleosynthesis therefore moves along a less direct channel. A silicon nucleus is attacked by very-high-energy photons, whose aggressivity has been fostered by the extreme temperature. Protons, neutrons and helium nuclei are torn off and grafted onto intact silicon nuclei to engender iron and its neighbours (see Appendix 3). But the lord of all elements is incarcerated at the heart of its kingdom. Indeed, it will not be enthroned. Its fate is to be neutronised, that is, transformed into neutrons by a process that we shall describe shortly.

It is no surprise that core silicon fusion should stumble on reaching iron when we remember that the latter is the most stable nucleus in nature. For this

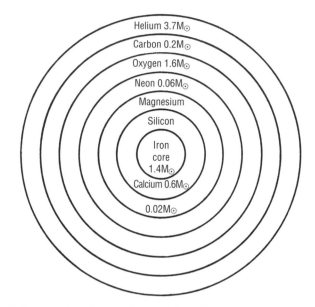

Fig. 7.3. Cross-section of a massive star. The onion-skin structure is specific to massive stars at the end of their existence.

simple reason, fusion reactions likely to produce heavier and more complex nuclei would absorb energy rather than releasing it. This would have a quite disastrous effect for stars!

Thus, when an iron core develops, reactions capable of generating energy come to an end and the star loses its only means of resisting against gravitational collapse. Disaster is indeed imminent.

But in its final hour, the star will never have been so beautiful within. If someone could only see through the opaque layers of the red supergiant, they could admire the whole range of nucleosynthesis reactions playing in symphony, one within the other, from silicon at the centre, through to hydrogen on the outskirts. And yet nothing on the healthy pink of the star's blushing countenance would lead us to suspect this nuclear apotheosis. At this point, the innocent star is about to abandon its blissful stationary state and undergo the most radical of its transfigurations.

Most of the elements propitious for the creation of life have now been put together. It remains only to share them with the rest of the Universe, to sow them across interstellar space. Explosion will accomplish this final task.

Explosion

Having painted this portrait of the inner beauty of the dying red star in all its splendour, the time has come to enter a little further into the physics of exploding massive stars. For the composition of the sky and the world so depend on it.

Let us sum up the main argument. The accumulation of iron and its weight category in the core of bountiful stars leads directly to catastrophic conditions insofar as nuclear reactions can no longer supply energy. The stars are no longer able to generate heat and pressure, and hence lose their stabilising mechanism. They are the victims of a genuine heart failure leading to a very fast rise in temperature.

In fact the temperature increases so quickly in the central regions that iron itself is smashed apart by ravaging photons. It is photodisintegrated into helium and this in turn is shattered into protons and neutrons. The density is such that the protons, usually so placid, swallow electrons to become neutrons, spitting out neutrinos like pips:

$$^{56}\text{Fe} + \gamma \longrightarrow 14\ ^4\text{He},$$
$$^4\text{He} + \gamma \longrightarrow 2p + 2n,$$
$$p + e \longrightarrow n + \nu.$$

The deep heart of the star, containing a mass of between 1.4 and 2 $M_\odot$, is thus neutralised, or rather, neutronised. The star is on its way to becoming a neutron star.

The phenomena of photodisintegration and electron capture deprive the star of support from photon and electron pressure. The former drain their energy in shattering iron nuclei, whilst the latter are purely and simply eliminated from the game. Under gravity's tyranny, the core suddenly implodes.

During collapse, which lasts only a split second, the temperature goes up to 10 billion K, as it was in the infant Universe when it was barely 1 second old. The density exceeds that of an atomic nucleus (10^{14} g cm^{-3}). Compressed like a spring, the matter then bounces back, for compression went just a little too far. This abrupt return to expansion gives rise to a shock wave that moves back out through the star.

Apart from this phenomenon, gravitational collapse has another important effect. The tremendously hot neutron star in the making emits a copious supply of thermal neutrinos and antineutrinos. These transfer some 10^{53} erg, that is, almost all the gravitational energy liberated by compaction of part of the original star into a neutron star with mass around 1.5 $M_\odot$ and radius 10 km.

The shock wave is soon exhausted for it encounters enough iron in its path to block its progress. Its energy is consumed photodisintegrating iron nuclei and it is attenuated and fades out. But help is on its way. The neutrino army strikes out, ready to communicate its outward drive. For given the extreme density of the matter the neutrinos must pass through, these usually discreet

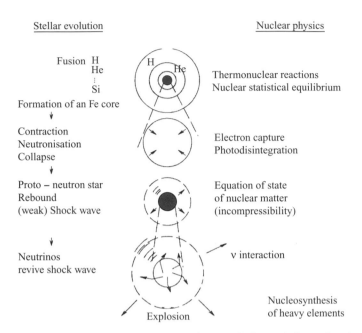

Stellar evolution   Nuclear physics

Fusion H
        He
        ⋮
        Si   Thermonuclear reactions
Formation of an Fe core   Nuclear statistical equilibrium

Contraction
Neutronisation   Electron capture
Collapse   Photodisintegration

Proto – neutron star   Equation of state
Rebound   of nuclear matter
(weak) Shock wave   (incompressibility)

Neutrinos   ν interaction
revive shock wave

                                              Nucleosynthesis
                                              of heavy elements
                Explosion

Fig. 7.4. Schematic description of an exploding star. Stellar evolution and nuclear physics are shown in parallel. In the first stage, hydrostatic fusion processes (H, He, C, O, Si) occur via thermonuclear reactions, leading finally to a core whose composition is governed by nuclear statistical equilibrium (see Appendix 3). Iron predominates in the core. The second stage is contraction, neutronisation and collapse, under the effects of electron capture and photodisintegration. In the third stage, the star counterattacks. As the matter within it has reached its maximum level of compressibility, it bounces back out, causing the star to expand suddenly. The abrupt movement triggers a shock wave. In the fourth stage the star explodes. Neutrinos revive the shock wave which exhausts itself by photodisintegrating iron nuclei lying in its path. Explosive nucleosynthesis of silicon, oxygen and carbon takes place in the wake of the shock wave. (Courtesy of Marcel Arnould, Université Libre, Brussels.)

and unsociable particles are forced to rub up against nuclei and pass on a part of their momentum. Not only do they thereby revive the shock wave, but they also take the credit for triggering the explosion of massive stars; at least this is the word in the best-informed astrophysical circles (Fig. 7.4).

Released in phenomenal quantities, only 1% of their energy need be communicated to the envelope in order to shatter it to smithereens. The revived shock wave reignites nuclear reactions in its wake and modifies the deep isotopic composition of the star (Fig. 7.5).

Implosion and explosion, Thanatos, the neutron star, and Eros, the supernova, seem to join their eternal forces, using invisible neutrinos as mediators.

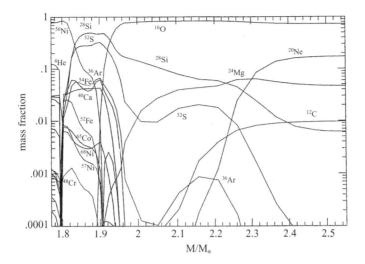

Fig. 7.5. Detailed isotopic profile of the central part of a 25-$M_\odot$ star after explosion. The graph shows the dependence of composition on depth in the star (radial profile). However, the scale on the horizontal axis is not given in terms of the radius, for this would make the diagram quite unreadable, the core being so drastically compressed. The scale is in fact graduated in included mass. The vertical axis gives the mass fraction of various nuclear species, that is, the number of grams of the relevant species per gram of stellar matter. The matter ejected by a 25-$M_\odot$ star (a typical type II supernova) exhibits a layered isotopic composition inherited from its complete past history, but modified deep down by passage of the shock wave. The latter literally volatilises the star. Only the innermost region of the supernova is shown (0.7 $M_\odot$). The collapse and rebound of the core thus engender a shock wave that sweeps through the star, heating up matter as it passes and detaching it from the collapsed core. (From Thielemann *et al.* 2001.)

Great liquid detectors like IMB (Irvine–Michigan–Brookhaven) and Kamioka, crouching underground to spy out solar neutrinos and catch protons in the act of decay, placidly recorded the neutrino signal of the Magellanic supernova in the form of a handful of blue flashes. Although the message had taken 170 000 years to reach us due to the great distance, it arrived just at the right moment, when human beings had developed sufficient physical models and computers up to the task of reconstructing stellar evolution by calculation. The message was late, but it was precise. The neutrinos arrived a few hours before the visible light signal and in the expected numbers, thus providing a magnificent confirmation of the scenario elaborated by explosive astrophysics. Other pleasant surprises were to follow, this time concerning the light curve of the wonderful supernova and its gamma-ray emission.

Proof of the explosive and radioactive origin of iron

The supernova of the century and indeed of modern times, SN 1987A was quite definitely of the gravitational type, as opposed to the thermonuclear species which we shall discuss shortly. In terms of its mass (20 $M_\odot$), the progenitor star, Sanduleak 92202, was not very different from Rigel, the beautiful blue star in the constellation of Orion. We know this from photographs taken before the cataclysm. Most stars explode when scarlet and distended, so much so that red has been taken as a sign of stellar senility and blue of youth. SN 1987A was an exception for it exploded whilst in blue attire (Fig. 7.6). Questions are still being asked about this blue death.[1]

At this stage in the investigation, the best hypothesis is that the late blueness is due to a combination of two phenomena:

- the metal deficiency of the Large Magellanic Cloud in which the star was born;
- the loss of 2–3 $M_\odot$, carried off by stellar winds before the explosion and appearing today in the form of circumstellar rings, greatly enhancing the beauty of the object.

Whatever caused this stunning blue sheen, it represented a unique opportunity to test the theory of how massive stars explode and how nucleosynthesis takes place within the explosion. This theory predicted that isotopes of mass 44, 56 and 57 would be produced by the sudden, explosive grafting of alpha particles (helium nuclei) and protons onto silicon nuclei (see Appendix 3). They would be synthesised in their radioactive forms, nickel-56, nickel-57 and titanium-44, in that order of importance (see Table 7.1). After a suitable series of decays, these sparsely scattered nuclei in the supernova debris would arrive at their stable forms, iron-56, iron-57 and calcium-44.

Such radioactivity could not go unnoticed. Indeed, radiation from the exploded star was to carry its unmistakable signature. In order to realise the great importance of this comment, we must now leave theoretical limbo and refer to observation, so true, so beautiful and so irrefutable. Let us then rewrite this chapter in more concrete terms.

When Cortez asked the Aztecs where they obtained the iron in their daggers, they pointed to the sky. Astrophysicists also suspected this to be the case and a startling confirmation was supplied by the 1987 exploding star, the long-awaited

---

[1] I would like to mention Richard Schaeffer of the Theoretical Physics Department at the CEA in Orme des Merisiers, France, and Robert Mochkovitch of the Institute of Astrophysics in Paris, with whom I calculated the light curve of an undressed star that looked so like the precursor of SN 1987A.

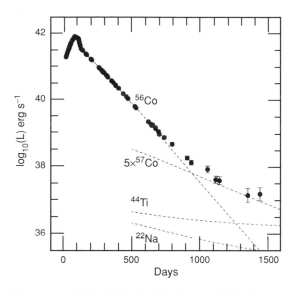

Fig. 7.6. Light curve of SN 1987A. On 27 February 1987, Shelton and Jones announced the discovery of a supernova in the Large Magellanic Cloud. It was the brightest observed since the one seen by Kepler in 1604, and the first that could be examined across the whole electromagnetic spectrum. It was also the first to be detected through its neutrino flux. Being relatively nearby (only 170 000 light-years), it represented a unique opportunity to observe a supernova in all its detail, and this with a variety of different detection techniques. After the first 200 days, the fall in the supernova light curve follows the radioactive decay of ^{56}Co (half-life 77 days), father of ^{56}Fe and son of ^{56}Ni. Light from the supernova is then maintained over longer periods by the decay of ^{57}Co and ^{44}Ti, making it a purely radioactive object. Energy deposited by radionuclides (dotted lines) corresponds to initial amounts of ^{56}Ni, ^{57}Ni and ^{44}Ti equal to 0.075, 0.009 and 0.0001 $M_\odot$, respectively.

supernova. It had been 400 years since Tycho Brahe, the master, and Johannes Kepler, the pupil, had both witnessed the appearance of a supernova, in 1572 and 1604, respectively. Ironically, the telescope was only developed in 1609. For four long centuries, we were then deprived of nearby explosions. Only in 1987 did a supernova appear that was visible to the naked eye, although not from the northern hemisphere. But when the Large Magellanic Cloud opened its box of treasures, what emotion there was before this splendour!

The 'new' star attracted the interest of astronomers in every guise, scrutinising the object in the visible and the invisible. Its light was collected by the very best telescopes in the southern hemisphere and dissected by spectrometers, whilst X rays and gamma rays were captured by sensitive devices carried aboard satellites.

Table 7.1. *Radioactive isotopes generating gamma rays, calculated from theoretical nucleosynthesis models*

Isotope	Mean lifetime[a]	Line energy, keV	Typical production, $M_\odot$				
			Wolf–Rayet stars	SNIa	SNIb/c	SNII	Nova
^{57}Ni	2.14 d	1378		0.02	0.005	0.005	
^{56}Ni	8.5 d	158/812		0.5	0.1	0.1	
^{7}Be	77 d	478			10^{-7}	5×10^{-7}	5×10^{-11}
^{56}Co	112 d	847/1238		0.5	0.1	0.1	
^{57}Co	392 d	122		0.02	0.005	0.005	
^{22}Na	3.76 y	1275		10^{-6}		10^{-6}	6×10^{-9}
^{44}Ti	87 y	1157		10^{-5}	5×10^{-5}	5×10^{-5}	
^{26}Al	10^{6} y	1809	10^{-4}		5×10^{-5}	5×10^{-5}	10^{-8}

[a] Time in days (d) or years (y).

It was soon noticed, to the great satisfaction of theoreticians, that the brightness of the object was falling at the rate implied by the decay of cobalt-56 (with a mean lifetime of 111 days or a half-life of 77 days), precisely as predicted. This decay was considered to be responsible for the supernova's light emissions at that stage. It was then replaced by the decay of cobalt-57, which in good time probably handed over to titanium-44.

It could thus be claimed that radioactivity is the source of light emissions from the 1987 supernova, and indeed of any similar supernova resulting from heavyweight star explosions. To crown it all, considerably strengthening the case for this argument, the Solar Maximum Mission satellite indicated that, six months after the visual appearance of the object, gamma photons of a certain very precise energy had begun to escape from the debris of the explosion. Because the ejecta expand and dilute, matter from the exploded star had just become transparent to photons produced by radioactive decay of cobalt-56, son of nickel-56 and father of iron-56.

This daring prediction was thereby confirmed: iron, lord of nuclear creation, the most robust of all atomic nuclei, is not created as iron, but as radioactive nickel (see Appendix 3).

And so it is that light from the supernova of modern times has come to reassure humankind of its ability to understand how the elements are synthesised in stars. At the same time, the fog of speculation has lifted from the theory of stellar evolution, culmination of astrophysical thinking, appointing it to the rank of a well-established science, based on a firm observational foundation.

Thermonuclear supernovas

Encouraged by this success, we begin to dream of explaining all the sky's cataclysms, from supernovas of every denomination, to hypernovas and gamma bursts. Remaining for the moment with supernovas, there appear to be two varieties. The first, described above, arises from core collapse in massive stars and leads to neutron stars and maybe black holes in some cases. These neutron stars then cool by emitting intense neutrino fluxes. The second type leaves no compact object behind and emits no neutrinos. However, these stars require a white dwarf closely accompanied by another object that will administer the kiss of death. The first type are called gravitational-collapse supernovas whilst the second belong to the class of thermonuclear supernovas, for reasons related to the cause of explosion.

This distinction replaces the outmoded classification into types II and I based on spectroscopic characteristics, in particular, the appearance of hydrogen lines for type II, or not for type I. The notation SNIa (for supernova type Ia) has nevertheless stuck for those beautiful supernovas whose light curve has certain well-specified inflections and in which hydrogen is striking by its absence. Likewise, SNII is still used to denote supernovas whose spectra are adorned with hydrogen lines. (Supernovas classed as Ib and Ic are in fact of the gravitational-collapse type. Their hydrogen envelope has been previously evaporated or expropriated by a companion star. These supernovas thus explode in basically the same way as the SNII.)

Thermonuclear supernovas are close in operation to bombs of the same name. Their brutal and complex physics involves matter exchange between two coupled stars, one of which is a white dwarf (Fig. 7.7). The trade-off is controlled by gravity in rather opaque conditions that render the whole process difficult to discern.

Among the various foul scenarios that lead to death by overdose, two stand out, with the common feature that they involve the same victim: the white dwarf. In the first, the companion is an authentic star like the Sun or a red giant. In the second, much less probable, it is another white dwarf.

Depending on the corpulence of its dancing partner, and how closely they hold one another, the white dwarf can increase or decrease its suction, thereby modifying the time elapsed before it explodes. Let us go to the stars' ball and, at the height of indiscretion, follow one of these scandalous couples, comprising one dead partner and one living. Resurrection of the corpse results in a new death which this time is final, scattering the remains over a wide area, if indeed death can be equated with total dispersion of atoms.

More seriously, it is generally agreed that most SNIa events result from the explosion of a white dwarf which is forced over the critical mass for such an

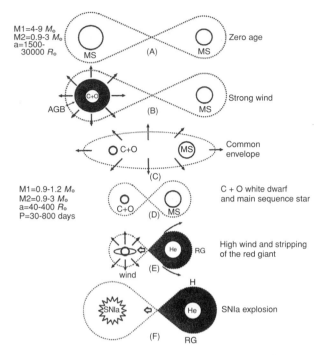

Fig. 7.7. Stellar duo. The presence of a companion star can considerably perturb a star's evolution. Hence, mass transfer by accretion transforms a rather dull white dwarf into an erupting nova or a type Ia supernova. As an example, let us follow the life of a star with mass between 4 and 9 $M_\odot$ and its little sister star with mass between 0.9 and 3 $M_\odot$, separated by a distance of between 1500 and 30 000 $R_\odot$ (where $R_\odot$ is the solar radius). In childhood, the system is calm. The big star evolves more quickly than the small one, however, a universal feature of stellar evolution. It soon becomes an asymptotic giant, sweeping the companion star with its winds, and then a white dwarf. The oxygen- and carbon-built white dwarf shares an envelope with its partner and together they evolve beneath this cloak as one and the same star. The result is a pair comprising a white dwarf with mass between 0.9 and 1.2 $M_\odot$ and a normal star with mass between 0.9 and 3 $M_\odot$, still evolving on the main sequence. The two components are separated by a distance of some 40–400 $R_\odot$, corresponding to a period of revolution of 30–800 days. The second star swells up and becomes a red giant. This is a boon for the dwarf. It captures the matter so generously donated. However, it cannot absorb it! A tremendous wind is generated and, in the end, a cataclysmic explosion ensues. (After Nomoto *et al.* 2001.)

object by adjunction of matter from some external source. This critical mass is called the Chandrasekhar mass, equal to 1.4 $M_\odot$. Once exceeded, the white dwarf disintegrates under the effects of thermonuclear fusion of its carbon and oxygen.

The precise manner in which it self-destructs depends on its own mass and that of its companion, but also on the nature of the latter and the distance between them. The detailed mechanism of the explosion remains somewhat of a mystery. This is because there seems to be some diversity within the SNIa class of events. As things are going at present, a new division will soon be necessary (see Appendix 2).

For the time being, we shall rejoice in the fact that SNIa explosions seem to have been largely understood, at least in outline, as suggested by the following three arguments:

- no neutron stars appear amongst the remains of past SNIa events;
- there is a certain level of uniformity in their behaviour which would follow in the final reckoning because white dwarfs are all much alike;
- there is good agreement with observed optical spectra.

Apart from these three facts, nuclear astrophysicists take pains to point out that the rate at which the luminosities of SNIa events decline, once beyond the maximum, is commensurable with the decay of radioactive cobalt-56, son of nickel-56, atomic nucleus of noble lineage as we know. This is a common factor with gravitational collapse supernovas. SNIa light curves are explained through the hypothesis that half a solar mass of nickel-56 is produced when one of these white dwarfs explodes.

This synthesis proceeds by total incineration of carbon and oxygen. The turbulent nuclear combustion occurs in such high density conditions that the matter is afflicted by quantum degeneracy. It need come as no surprise that white dwarfs suffer from a certain lack of flexibility, at such densities. The matter making them up does not behave like a perfect gas, ajdusting its structure to the slightest accident. Rigor mortis has already set in. If a nuclear reaction gets under way, the white dwarf is condemned to burn at the stake. Cremation is practically instantaneous and the whole thing goes up in a radioactive nickel smoke.

Nuclear incineration of the gluttonous white dwarf thus synthesises a considerable amount of nickel-56 (0.5–0.6 $M_\odot$) and radioactivity left over after its departure makes it shine with dazzling brilliance, like some lavish stellar requiem. This radioactive isotope is the source of the exceptional luminosity of type Ia supernovas, both in the optical region, as has already been observed, and in gamma rays, as yet only a prediction.

The European satellite INTEGRAL (Fig. 7.8) will soon be ready to seize upon any object of this type having the good taste to explode within 45 million light-years of it whilst it happens to be in operation.

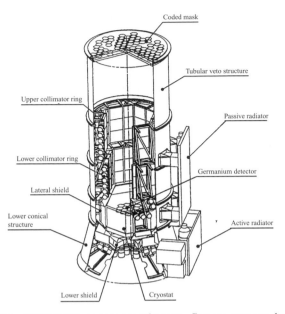

Fig. 7.8. The INTEGRAL gamma-ray telescope. Gamma rays are detected by means of a coded mask, which also allows the computer to reconstruct the direction of incidence (Paul 1998). The instrument is designed on the basis of a compact hexagonal geometry. The distance between the coded mask and the detector has been chosen to obtain a decent spatial resolution and field of view. The main features are: (1) a detection plane made up of 19 exceptionally pure germanium crystals, (2) an active shield made of bismuth germanate, designed to avoid unwanted detections due to cosmic-ray protons, and (3) a coded mask. Detectors are cooled by an active cryogenic system. High-precision energy measurements are afforded by the use of germanium detectors. A semiconducting material, germanium becomes a true conductor when gamma rays release electrons or set them in motion by the photoelectric or Compton effects. The INTEGRAL spectrometer has the following characteristics: energy range 20 keV to 8 MeV, energy resolution 2 keV at 1 MeV, accuracy of angular measurements 2° for point sources, field of view 16°, narrow line sensitivity 3×10^{-6} photon cm^{-2} s at 1 MeV, timing accuracy 100 µs, detector area 500 cm^2.

However, not all pairs of stars will meet such a cataclysmic end. The proof stems from the existence of novas. These also arise from the tempestuous love affairs between a white dwarf and a healthy star. The difference is that their love burns in a more reasonable manner, ejecting only a small portion of their envelope at a time (roughly 10^{-4} $M_\odot$). These ejecta are nevertheless loaded like galleons with radioactive isotopes such as beryllium-7 and sodium-22. One day it is hoped that their signature will be detected by the great INTEGRAL observatory.

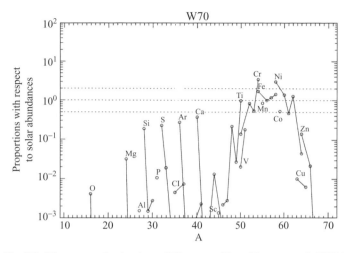

Fig. 7.9. Element production in an SNIa event. (From Nomoto *et al.* 1997.)

In praise of supernovas

Supernovas feature among the most beautiful objects in astronomy, whatever class they may belong to. In addition, they are a genuine driving force in galactic evolution. They are mean neither with their energy (10^{51} erg) nor with newly wrought matter (2 $M_\odot$ of oxygen and 0.6 $M_\odot$ of iron for SNIa and SNII, respectively). The two types of supernova do not produce the elements in the same proportions and they do not occur at the same rate (one thermonuclear for five gravitational collapse). All this is reflected in the galactic evolution of oxygen and iron (see Appendix 5). It is interesting to note that, if pre-explosion iron were not transmuted in neutron stars, but instead were somehow scattered into space, its production would rival that of oxygen and the two elements would have comparable abundances. The consequences would be considerable.

Gravitational-collapse supernovas are efficient producers of a whole range of elements between carbon and calcium, oxygen being the most abundant (see Appendix 4). At the same time, their thermonuclear siblings are more than generous with iron and neighbouring elements (Fig. 7.9). It seems that Solar System abundances can be explained, at least qualitatively, by a well-dosed cocktail of ingredients from the two sources.

Apart from the fact that they represent the most productive explosions in the sky since the Big Bang, they can also be used as a waymark across space. The depths of the cosmos are lit up by light from SNIa events. At its high point, just one such event shines like a billion suns, on a par with a small galaxy. By virtue of their extraordinary brightness, they can be seen billions of light-years away. By their brightness and similarity, they have become a precious tool for astronomy. Indeed, it is hoped that they will allow us to measure the Hubble parameter, which quantifies the rate of expansion of the Universe, both in the present and in the past. It will then be possible to estimate the amount of quintessence needed in a good recipe for the Universe (see Appendix 1). Observational data gathered so far would suggest that quintessence dominates at the current epoch, but that this was not always the case. If it had been, the sky would be quite empty of galaxies.

Ferrous destiny

This is very fine and beautiful, but when are we going to explain why hydrogen rather than iron heads the list of universal abundances? We have already whispered the answer, but rather discreetly. Having determined how iron is produced, the time has come to deal with this question.

According to the law of the jungle, whereby the strongest prevail, the element whose nucleus contains 26 protons and 30 neutrons stands out in principle as the lord of all nuclear creation. However, this rudimentary microscopic Darwinism is belied by at least two arguments.

The first is a historical fact of prime importance, namely, the low percentage of residual gas in the Milky Way. It is true that iron comes from the stars, and that these seem to advocate its claim to the throne, but star formation itself is on the decline through lack of gas. For this reason, the iron content of the Galaxy levelled off several billion years ago, having reached what appears to be its maximum (see Appendix 5).

An even deeper reason explains why iron does not reign over the other elements in the Galaxy, despite its remarkable toughness. The very high temperatures required to build up complex isotopes can only occur in the deepest stellar confines, from which matter has great difficulty in extracting itself. The fact that elements are created somewhere does not mean that we will find them in nature. They must also be able to escape from the stellar crucible. Relegated to the deepest quarters of massive stars, iron and its kin have great difficulty in tearing themselves away. Only an explosion can simultaneously create them and free them, and even then gravitational collapse supernovas confiscate the pre-explosion iron in their bowels for subsequent digestion. The incarceration of the most stable nuclei in stellar corpses (neutron stars or black holes) is one reason for their rarity. The king and his court are prisoners, locked up in the dungeons where they transmute into neutrons and who knows what else.

However, the entombment of iron is only perpetrated by gravitational-collapse supernovas. Their thermonuclear counterparts are more liberal and, one might say, more final, for they leave behind no corpse, no bones, and no scrap iron. They owe this propensity for total destruction to the rigidity and fragility of the exploding body, the white dwarf, a porcelain ornament that is sure to break when it falls. But thermonuclear supernovas, though lavish providers of iron, are rare. Very special conditions must be fulfilled for these explosions to occur.

For all these reasons, iron will always maintain a modest presence and we are saved from the terrifying vision of a completely metallic Universe. So, dear Aristotle, we no longer stand before the evidence like bats in daylight.

Hypernovas and gamma bursts

Gigantic blasts of very-high-energy photons are sometimes recorded by satellite-borne detectors. Such gamma bursts have long remained a mystery.

Over recent years, theoretical understanding of these monstrous outbursts has made some progress, thanks mainly to a rich crop of results obtained by the BATSE experiment on board the Gamma-Ray Observatory and the Dutch–Italian satellite Beppo-SAX. The bursts themselves, and the pale afterglows that succeed them in other wavelength ranges, can be described as resulting from interactions between the parts of a highly relativistic jet or between the surrounding medium and such a jet. Current models of this kind are poetically referred to as the cosmic or relativistic fireball. There is still a great deal of debate over the origin of the jet. However, it is generally agreed that it involves the formation of a black hole with mass between two and five times the solar mass, a disk around it, and rapid accretion of disk material by the black hole. Different theorists have different ideas about how the black hole may have formed and how much matter accretes and for how long. Opinions also vary on how the disk energy is extracted and converted into a beam of relativistic particles.

Hypernovas and gamma bursts are indeed the biggest explosions since the Big Bang, at least, the biggest to be observed by humankind. The energy involved exceeds conventional bounds by a wide margin. In gamma rays alone, some 10^{53} to 10^{54} erg are released, assuming emissions to be isotropic, that is, equally distributed in all directions. This is comparable with the energy carried off by neutrinos during the explosion of a massive star, like SN 1987A.

The immensity of this figure suggests that emissions may not in fact be om-nidirectional. The total energy output can be substantially reduced by assuming that gamma rays are channelled along a beam. This beam has to be remarkably narrow if the emitted energy is to remain within respectable limits.

This raises the question as to what fabulous engine might produce and guide these lavish gusts of gamma photons.[2]

Let us muse for a moment over some figures. The energy equivalent of the Sun's mass, calculated using the well-known relation $E = mc^2$, is around 2×10^{54} erg. The core of a massive star (some $2\,M_\odot$) contains twice this energy. The problem is thus to convert the mass energy of a body, or part of it, into gamma energy, extract it from the object during its collapse phase, and channel it into a narrow beam of gamma rays over a brief period of time. In other words, how could we translate from the physics of stars to the physics of beams?

Apart from its intensity, what is striking is also the brevity of the phenomenon, earning it the name of cataclysm. The key seems to be the collapse of the iron core of a massive star, but this time, one that is rotating. Carried along by the dance, it spins around, and so does its core, at such a rate that the rotational energy can reach a considerable value. In fact the rotational energy may even be on a par with the gravitational energy.

The faster the rotation, the easier it is to extract energy. To this end, we know of at least one mechanism, the Blandford–Znajek mechanism. Briefly, the principal suspect is a rapidly spinning black hole. When the rotation is abruptly slowed down, a prodigious quantity of energy is released. To a first approximation, a black hole swallows everything and gives back nothing at all, unless it is spinning. We thus appeal to rotation to resolve the enigma of the gamma-ray blasts.

Something must slow down the rotation of a black hole with mass greater than the solar mass (a Kerr black hole), causing it to stop spinning (and thus become a Schwarzschild black hole). This must happen within a few hundred seconds. The energy must be extracted and emitted in the form of intense gamma radiation for one or two seconds, so that it can be registered as a gamma burst by satellites on permanent watch around the Earth.

Speculation has been rife since a gamma burst was found to coincide with an extragalactic supernova with registration plate SN 1998bw. It should be said that this supernova is one of the brightest ever detected, in both the optical and the radio regions. Indeed, its spectrum is somewhat enigmatic. Its general appearance corresponds to an abnormally high expansion rate. The fingerprint of hydrogen is not to be found on the scene, either as an emission line or an absorption line. The event is not therefore of type II.

---

[2] Useful references are the PhD thesis of Frédéric Daigne (Institute of Astrophysics, Paris, 1999) and the work of Robert Mochkovitch.

Detailed analysis of observations leads to the following breakdown for SN 1998bw:

mass of exploded star	13.8 $M_\odot$
mass of radioactive nickel ejected	0.7 $M_\odot$
mass of ejected matter	10.9 $M_\odot$
energy of explosion	3×10^{52} erg
mass of compact remnant	2.9 $M_\odot$
mass of precursor star	30–40$M_\odot$.

It would be tempting to call this event a hypernova, if the term had been officially accepted, and to associate a black hole with the surviving compact remnant.

In this case, we must examine the effect of an asymmetrical explosion on the final result of the self-destruction of a very massive black hole, that is, the supernova remnant and black hole. Such asymmetrical supernovas may explain a fair number of gamma bursts, which have remained a deep mystery up to now. Beams and jets are more than ever in the news in astrophysics.

Collapse and accretion are other key words which no one dabbling in astrophysics can ignore today. To accrete means to attract and appropriate matter from an external source. Accretion always goes with gravitation, the attraction of matter by matter. In every couple, there is a winner and a loser. The winner is the one who, by superior force, appropriates the substance of the other. A high mass and small radius are qualifying features. So we have the burglar and the burgled. In this game, black holes are the champions, the most accomplished plunderers in the sky, but it is the desperate cry of their victims that gives them away. Matter sends out a distress signal before being absorbed forever, and this strident vociferation is best heard in the X-ray or even gamma-ray region.

A wide range of phenomena is possible when a black hole absorbs matter very quickly from the disk around it after incomplete explosion of a rapidly rotating massive supernova. In the most extreme case, according to Stan Woosley (University of Santa Cruz), the shock wave resulting from the central explosion is unable to shatter the star (composed mainly of helium) and a black hole forms with an accretion disk around it. A high-intensity gamma burst then results.

We may wonder what effect these gigantic explosions and accompanying radiation may have on the galactic environment of the hypernova. Such an energy release could only produce holes, hollow shells in the distribution of interstellar matter. After 1 million years, these cavities reach a radius of 150 to 500 light-years, according to calculation. The initial excesses are followed by a calmer expansion at speeds below 10 km s^{-1}.

Hypernovas

Neutrinos are usually evoked as detonators, triggering or driving the explosion of massive stars. The gravitational energy freed by core collapse of a massive star, some 10^{53} erg, is mainly evacuated by neutrinos. These pour out in such inordinate amounts that if just 1% of their energy is communicated to matter in the envelope, the whole thing will be blasted out of existence.

Over the past thirty years, this idea has beclouded research in the field. After examining certain events more closely, the explosion community has begun to wake from a long dogmatic slumber. Previously, the cosmic firework-makers loved to see round supernovas, as beautiful as fruit, whilst in reality it seems that they may be pierced by jets and beams, like balls of wool with knitting needles through them.

The conceptually simplest way of forming a black hole at the heart of a massive star, thereby setting up the conditions of the hypernova model, is to begin by repudiating the traditional explosion model detonated by neutrinos. The iron core then collapses without remission in the space of one second. A black hole prospers, pulling down the rest of the stellar edifice. This may be a common occurrence for stars of 35 to 40 $M_\odot$. However, uncertainties remain concerning convection, mass loss and mixing due to rotation, not to mention the explosion mechanism itself.

If the star loses its hydrogen envelope in the process, and if the jet produced by accretion maintains its energy and remains focussed longer than the time required to cross the star (five to ten seconds), a vigorous gamma burst is produced. Otherwise, the result is a weaker burst, less tightly collimated, or an asymmetrical supernova.

However, according to Stan Woosley, there must be a whole range of masses in which a black hole is not immediately created, but only when a shock wave has blown the star apart. One would feel sure that the explosion had succeeded, and yet a certain fraction of the matter would fall back into the core, for it would have insufficient kinetic energy to resist the call of gravity from the central neutron star. The latter would be transformed into a black hole by the extra matter. This delayed delivery of a black hole may be much more common than the hasty birth described above.

As an example, let us consider the case of a 25 $M_\odot$ black hole. Let us examine it from a young age so as not to lose anything of its life story. On the main sequence, it maintains a significant stellar wind and loses mass. At the end of its life, it graces the sky as a red supergiant. It has an iron core of 1.9 $M_\odot$, a helium shell of 8 $M_\odot$ and a downy envelope weighing in at 6.6 $M_\odot$. Its total mass at death is 14.6 $M_\odot$ and it measures 8×10^{13} cm in radius. The star has enough specific angular momentum ($\sim 10^{17}$ cm^2 s^{-1}) at the equator to form an accretion disk around a black hole. With a disk and a black hole, all the ingredients are there for a great firework show.

SN 1998bw, which sounds something like the name of a Bach fugue, is not the only animal belonging to this area of the zoo. Another carries similar features in its spectrum, namely SN 1997ef. (Note that, given the regularity with which supernovas are discovered and recorded these days, their registration plate now carries two extra letters.) This was discovered on 25 November 1997 in an obscure spiral galaxy UGC 4107, situated some 170 million light-years from our own. Its spectrum is dominated by broad oxygen and iron lines. It harbours no trace of hydrogen, suggesting that it is a gravitational-collapse supernova that has lost its envelope under the effects of a strong stellar wind prior to explosion.

SN 1997ef is a very unusual supernova, to judge by its light curve and spectrum. Hydrogen is striking by its absence. Oxygen and iron absorption lines are abnormally broad.

The light curve and spectrum are well reproduced by numerically simulating the explosion of a bare oxygen and carbon core with mass 6 $M_\odot$. Since the width of the absorption lines is not thereby explained, models with higher mass and explosion energy have been explored. A convincing representation of the phenomenon has been obtained by pushing the core mass up to 10 $M_\odot$ and the explosion energy to 10^{52} erg. The mass of the C + O core corresponds to a total stellar mass of 30–35 $M_\odot$ at birth. The final compact residue of 2.4 $M_\odot$ is greater than the maximum allowed mass of a neutron star and must therefore be a black hole. The kinetic energy of the explosion quite clearly places SN 1997ef in the hypernova category.

The same analysis has been applied to several supernovas. It is found that the mass of radioactive nickel ejected increases with the mass of the progenitor star, with one exception, namely SN 1997D (see Fig. 7.10). This trend can be explained as follows. Stars starting out with a mass less than about 25 $M_\odot$, such as 1993J, 1994I and the celebrated SN 1987A, would lead to a neutron star and produce a mass of 0.08 ± 0.03 $M_\odot$ ^{56}Ni. It transpires that SN 1987A adds to its originality by producing a compact remnant that is borderline between a neutron star and a black hole. Stars of mass greater than 25 $M_\odot$ would form a black hole without further complication. Whether they become type II (ordinary) supernovas or hypernovas depends on the rotational state of the core of the collapse victim. In the case of 1997D, because of the high gravitational potential, the explosion energy is so weak that most of the nickel-56 falls back onto the compact object forming at its centre. This matter adds to the mass of the imploded core causing total collapse and transformation of the neutron star into a black hole. The core of this supernova would not appear to have the

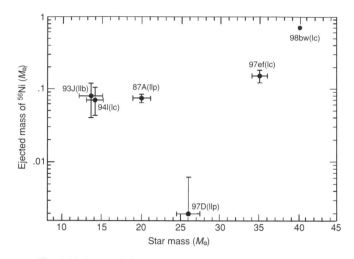

Fig. 7.10. Mass of nickel-56 (iron) ejected by high-mass stars.

angular momentum required to afford it hypernova status, for the progenitor star possessed a vast (hydrogen) envelope which somehow, by a kind of flywheel effect, slowed down the core rotation.

All this remains pure speculation and, in science, it is important to be cautious. I have nevertheless chosen to discuss the case of these very-high-energy explosions because it does illustrate some of the trends in contemporary astrophysics. In particular, it exemplifies the spectacular entry on stage of the black hole, which I personally regard with a dark eye.

One may wager, without risk of contradiction, that examples of this kind are going to accumulate. A third case of a hypernova (SN 1997cy) has already been revealed. We will soon know the amount of iron (nickel) engendered by supernovas and hypernovas of all masses. This is of the utmost importance for those like Elisabeth Vangioni-Flam who seek to reconstitute the history of this element in the Galaxy and beyond.

In the future, rapidly rotating carbon–oxygen cores will be most carefully examined by numerical astrophysicists, because the explosive energy of hypernovas may well be extracted from the resulting black hole, at least, if we are to believe Blandford and Znajek.

History of gold

The rise in temperature that follows in the wake of the shock wave triggers a series of nuclear reactions in the central regions of the star. It resuscitates nucleosynthesis of silicon and thereby produces iron that will not later be impounded, in contrast to the iron created in the core of the star before explosion. This iron is generated in the form of radioactive nickel-56 which gives itself away by the gamma rays it emits, as already discussed.

Explosive nucleosynthesis adds a few last trinkets to the abundance table, in particular, gold, platinum and uranium, through an ultimate nuclear process relating to neutron physics.

The key point in this respect is the large number of neutrons produced in the central region. Insofar as these nucleons carry no electric charge, they mix easily with the previously produced nuclear isotopes, including iron. They do not suffer the electrical barrier that so frustrates the fusion of nuclei, and ever more so as they occupy higher positions in the hierarchy. Neutron capture serves to enrich the range of nuclei that can be engendered by supernova activity.

This form of nucleosynthesis, known as the r process (or rapid process), is assumed to produce the more complex elements existing in nature, those lying well beyond iron, according to the following series of reactions:

$$^{56}\text{Fe} + \text{n} \longrightarrow {}^{57}\text{Fe} + \gamma,$$
$$^{57}\text{Fe} + \text{n} \longrightarrow {}^{58}\text{Fe} + \gamma,$$
$$^{58}\text{Fe} + \text{n} \longrightarrow {}^{59}\text{Fe} + \gamma,$$
$$\vdots$$
$$^{78}\text{Fe} + \text{n} \longrightarrow {}^{79}\text{Fe} + \gamma.$$

These reactions generate nuclei that are extremely rich in neutrons, in fact, much too rich to be stable. In this situation, radioactive β^- decays, a manifestation of the weak interaction, intervene to readjust the numbers N and Z, transforming excess neutrons into protons via the elementary reaction

$$\text{neutron} \longrightarrow \text{proton} + \text{electron} + \text{antineutrino}.$$

This results in a gradual transition towards higher atomic numbers Z.

The sequence of neutron captures followed by β decay produces heavier and heavier elements. It is the only known mechanism for producing gold, platinum, thorium and uranium. The r process is thus considered to close the cycle of nucleosynthesis.

In this way, the stars make gold. If some alchemist surfaces today from the distant past and asks you for the secret behind the manufacture of precious metals, why not give him the recipe! For good results, an iron nucleus must absorb about a hundred neutrons. Destroy any supernumerary iron (heat or photodisintegrate), add a generous dose of neutrons, and there you are!

Unfortunately, it is easier said than done. You need a cauldron, crucible, transmuting heat, ash, combustion, fire and light carried to an absolute extreme of violence. Celestial alchemy is an explosive business.

Bear the formula in mind, however. The ingredients are metals and neutrons. Neutrons! Here lies the whole difficulty of the matter, for free neutrons are unstable. How can they be liberated and made to react before they perish? What is needed is a source of neutrons, a neutron-rich nucleus that will let one of its neutrons slip out from the folds of its robe. Neutron sources so far identified are few and far between. Indeed, we have only carbon-13 and neon-22. These are the only ones capable of supplying neutrons to nuclear reactions. And we are still worlds away from working out a detailed scenario for the explosive production of gold.

History of lead

Just to reiterate what we have said, neutron capture is the only valid channel towards the extreme complexity of gold ($Z = 79$). Reactions involving charged particles are energetically unfavourable and moreover inhibited by insurmountable electrical barriers. Because of the strong electrical repulsion between heavy nuclei (which thus contain many protons), the classic thermonuclear fusion reactions are ineffective, and we are forced to accept the idea that nuclear species beyond iron are produced by a process other than thermonuclear fusion. This process is neutron capture.

We have seen that, globally speaking, nuclei above iron can be divided into two groups: those resulting from the s process and those originating in the r process. A third mechanism called the p process only affects a minority of nuclear isotopes, producing a few rare species rich in protons.

The term 's process' is an abbreviation for 'slow neutron capture process'. Here, capture is slow relative to the characteristic time for internal transformation of the neutron into a proton (radioactive β^- decay). Between two neutron captures, there is ample time for β decay to occur. The r process represents quite the opposite situation. Neutron capture is not interrupted by β decay.

As a nuclear reaction, the s process is relatively well understood, but the problem lies in identifying an astrophysical site for it and determining the relevant physical parameters, such as neutron flux, mean time separating two neutron captures, and temperature. It has been shown that the most propitious temperatures are those of helium fusion. Added to the fact that the surfaces of certain red giants are rich in s isotopes, such as radioactive technetium and barium, this observation confirms the idea that the s process may be related to helium fusion regions in stars.

More precisely, the stars in question do not belong to the main sequence, but rather to the asymptotic giant branch (or AGB, to use the astronomer's term, already mentioned several times in the above). These are red giants that began life with masses between three and eight times the mass of the Sun and which manage to produce genuine neutron bursts in an almost feverish manner. For this reason they constitute ideal sites for (slow) neutron capture by pre-existing iron. During these fitful episodes, stirred up by mighty blasts of heat referred to as thermal convulsions, neutrons are produced by the reaction

$$\text{helium} + \text{carbon-13} \longrightarrow \text{oxygen-16} + \text{neutron}.$$

Neutrons needed to develop the s process are essentially produced at temperatures of the order of 100 to 150 million K.

In fact, red giant nucleosynthesis is supposed to explain s elements with atomic mass greater than 100, that is, the heaviest among them. Lead and bismuth mark the end of the fabrication process for complex nuclei in giants. This is because further ingurgitation of neutrons by these elements results in unstable isotopes which transform back into lead and bismuth.

The neutron bursts take place in helium burning shells surrounding the inert carbon–oxygen core. The neutrons released here are grafted onto iron and its kin.

Massive stars, or rather their cores, in which helium is burning, are also candidates for producing s isotopes, but with masses less than 100 now. Neutrons are released via the reaction

$$\text{helium} + \text{neon-22} \longrightarrow \text{magnesium-25} + \text{neutron}.$$

In the competing rapid process, or r process, neutron fluxes are considerably greater. The classic r process is assumed to occur at extremely high neutron densities, in the range of 10^{20} or 10^{30} neutrons per cubic centimetre, and temperatures in the vicinity of 1 billion K. Given the intensity of irradiation, the number of neutrons absorbed, let us say by iron, reaches spectacular levels. The nuclear species formed in this way is very unstable, being far too rich in neutrons to remain a viable nucleus. It stabilises itself by transforming a large number of neutrons into protons one after the other, as we saw previously.

Hence, when neutron irradiation ceases, the exotic and extravagant nuclei undergo a chain of β decays to re-enter the valley of stability, that is, the region of stable nuclei which traces out a parabola in the (N, Z) plane, by successive transformation of neutrons into protons.

The r process can form a bridge between lead and the actinides above the nuclear instability zone. It is thus responsible for the production of long-lived actinides such as thorium-232, uranium-235, uranium-238 and plutonium-244 which are used to estimate the age of the Earth and the Galaxy.

In order to determine the maximum atomic mass produced in the r process, we must find the point when induced (destructive) fission enters into competition with (constructive) neutron capture on the path followed by the process across the (N, Z) map of the isotopes. This question requires calculation of the fission barrier far from the region of known nuclei, which is no simple matter. The possibility of producing mythical, superheavy, transuranium nuclei (around $Z = 114$ and $N = 184$) has not yet been demonstrated.

Up to now, the search for an astrophysical site that could sustain the r process has not brought much success, but it is certainly not for want of imagination. Mergers between two neutron stars or a neutron star and a black hole have even appeared on the list. Notwithstanding, the favourite potential site remains the supernova. However, despite a long inquiry into the matter, we are still unable to put forward a detailed mechanism to show how it would operate. Calculations with the r process in explosive conditions are notoriously difficult, but they are being pursued with courage and determination.

In fact, the quest to find an astrophysical site for the r process is in full swing. It is being helped along by the rapid accumulation of observational data concerning surface abundances in ageing stars gravitating in the galactic halo. Such analyses aim to correlate observed abundances of typical r or s isotopes with stellar metallicities. It has thereby been discovered that the slow process comes later than the rapid process.

Sharing the stellar treasure

The nuclear treasures accumulated by stars are now complete and we may draw up an inventory. But treasure is only treasure if it can be shared. What would be the point of the stars' unflagging labours if their fruits could not be redistributed? Nature has resolved the distribution problem by explosion or wind ejection. The star throws everything it can overboard before it goes down. In the sky, winds, blasts and deflagrations are entirely beneficial.

The explosion of a supernova is a very happy event. It results in the spherical propulsion of matter into the interstellar medium, matter that has been simmering over millions of years, spiced up in the final moments by a little explosion and radioactivity. In the medium that lies between the stars, temperatures and densities are much lower than in stellar objects themselves. The supernova matter is diluted and cools down. Nuclei in the expelled material capture electrons to form various atoms and molecules. The cycle

$$\text{cloud} \longrightarrow \text{star} \longrightarrow \text{cloud}$$

continues to make its way until the word 'cloud' can no longer find substance.

To begin with, the gravitational force condenses matter to form a new generation of astronomical bodies of all sizes, from massive stars with an explosive vocation, through stable and long-lasting solar-type stars to planets, meteorites and cosmic dust. Each star then follows its predestined track, not without making its contribution to the cosmic hamper. In this way, successive star generations gradually enhance the galactic humus in heavy elements. Clouds serve the double purpose of receiving the ashes of defunct stars and providing construction material for new stars. So goes the Galaxy.

But all this cannot happen without losses along the way. Stellar corpses and collapsed cores (white dwarfs, neutron stars and black holes) are permanently removed from the great flow of nuclear evolution. It is as though their substance has been confiscated, so that it can no longer take part in the ebb and flow of matter, entering the stars in one form and re-emerging in another. Almost all elements required for life are now present.

At this point in the presentation, our picture of stellar evolution and the nucleosynthesis of naturally occurring chemical elements is almost complete. Many refinements would be required to give a fuller view. In fact, many pieces of the jigsaw are still missing or inadequately understood, in particular with regard to the r process. Despite all this, we have tried to convince the reader that the nucleosynthesis model constitutes a fundamental opus of the human intellect.

It is not therefore to the planets that we should associate the elements: iron with Mars, lead with Saturn, mercury with Mercury. It is indeed the stars that have nurtured them. Some stars make carbon, others gold. Thermonuclear combustion modifies the composition of the hottest regions within stars. Each star is responsible for the confection and distribution of a particular batch of atoms, apart from hydrogen and a large part of the helium in the Universe which were synthesised in the Big Bang, and the lightweight trio lithium, beryllium and boron.

Explosive events like the Big Bang and supernovas are the professionals in the nucleosynthesis game. They are the great dispensers and generous donators of atomic nuclei in the Universe. The quantity and simplicity of nuclear species created by the Big Bang – hydrogen and helium – can only be balanced by the quality, diversity and refinement of species produced in supernovas, including 90 atomic types from carbon to uranium.

Stellar winds and planetary nebulas also play an important role in the chemical economy of our Galaxy, as they probably do in all the others. In particular, they enhance levels of nitrogen, carbon and heavy elements beyond iron (by the s process). For the main part, newly made elements are produced and launched into circulation by the last gasp of light stars (generating planetary nebulas), stellar winds and supernova explosions.

In the last analysis, the diversity of atoms reduces to a well-ordered numerical sequence. Furthermore, this order is significant and the final picture, summing up the origin of the elements, will be quite indispensable to anyone who sincerely seeks out the truth about matter. Such is the genuine and communicable fruit of our inquiry, to be shared by all. We have identified the creation, not in a metaphorical way, but literally, through the constitutional algebra of the Big Bang, cosmic radiation and stars. If it is true to say that '– our fortune or misfortune – depend on one thing, namely the quality of the subject to which we bind ourselves in love' (Spinoza), then we star-lovers have nothing to fear as we gaze upon the sky. In the institutes of astrophysics, a new generation has been brought up to understand it.

8

Ancient stars in the galactic halo

Glossary

damped Lyman alpha (DLA) systems thick intergalactic clouds absorbing quasar light

Clues to the chemical evolution of the Galaxy

The evolution of the composition of matter can be traced back through the various ages of the Galaxy by systematically examining surface abundances over a vary large population of stars by means of spectroscopic analysis (Table 8.1). One is particularly interested in elements observed in the spectra of ancient suns in the galactic halo. These little stars, the oldest we know of, are still shining valiantly today, boasting their exceptional longevity (Fig. 8.1).

Let us now describe the method used. The most accessible elements are those possessing clear lines in the optical spectra of these fossilised objects. In contrast, certain elements like neon and argon are not determined in these stars, whether they be dwarfs or giants. In their normal state, the noble gases produce no optical emission.

Families that lend themselves best to this evolutionary analysis are:

- the light nuclei Li, Be and B;
- the α nuclei, i.e. multiples of the helium nucleus, such as Mg, Si, S and Ca;
- nuclei around the iron peak, viz. Sc, Cr, Mn, Fe, Co, Ni, Cu and Zn;
- heavy s and r isotopes like Sr, Y, Ba and Eu.

Among these, iron is relatively easy to measure and serves as a reference, as a metallicity index, and thus as an indicator of the degree of evolution. Indeed, it is common practice in astronomy to treat the terms iron content (Fe/H) and metallicity (Z) as synonymous. Solar metallicity is denoted $Z_\odot$.

171

Table 8.1. *Lifetimes of stars with different masses and metallicities; Z denotes metallicity here and* M *is in units of solar mass*

M	Z = 0.0004	Z = 0.004	Z = 0.008	Z = 0.02	Z = 0.05
0.6	4.28×10^{10}	5.35×10^{10}	6.47×10^{10}	7.92×10^{10}	7.18×10^{10}
0.7	2.37×10^{10}	2.95×10^{10}	3.54×10^{10}	4.45×10^{10}	4.00×10^{10}
0.8	1.41×10^{10}	1.73×10^{10}	2.09×10^{10}	2.61×10^{10}	2.33×10^{10}
0.9	8.97×10^{9}	1.09×10^{10}	1.30×10^{10}	1.59×10^{10}	1.42×10^{10}
1.0	6.03×10^{9}	7.13×10^{9}	8.46×10^{9}	1.03×10^{10}	8.88×10^{9}
1.1	4.23×10^{9}	4.93×10^{9}	5.72×10^{9}	6.89×10^{9}	5.95×10^{9}
1.2	3.08×10^{9}	3.52×10^{9}	4.12×10^{9}	4.73×10^{9}	4.39×10^{9}
1.3	2.34×10^{9}	2.64×10^{9}	2.92×10^{9}	3.59×10^{9}	3.37×10^{9}
1.4	1.92×10^{9}	2.39×10^{9}	2.36×10^{9}	2.87×10^{9}	3.10×10^{9}
1.5	1.66×10^{9}	1.95×10^{9}	2.18×10^{9}	2.64×10^{9}	2.51×10^{9}
1.6	1.39×10^{9}	1.63×10^{9}	1.82×10^{9}	2.18×10^{9}	2.06×10^{9}
1.7	1.18×10^{9}	1.28×10^{9}	1.58×10^{9}	1.84×10^{9}	1.76×10^{9}
1.8	1.11×10^{9}	1.25×10^{9}	1.41×10^{9}	1.59×10^{9}	1.51×10^{9}
1.9	9.66×10^{8}	1.23×10^{9}	1.25×10^{9}	1.38×10^{9}	1.34×10^{9}
2.0	8.33×10^{8}	1.08×10^{9}	1.23×10^{9}	1.21×10^{9}	1.24×10^{9}
2.5	4.64×10^{8}	5.98×10^{8}	6.86×10^{8}	7.64×10^{8}	6.58×10^{8}
3	3.03×10^{8}	3.67×10^{8}	4.12×10^{8}	4.56×10^{8}	3.81×10^{8}
4	1.61×10^{8}	1.82×10^{8}	1.93×10^{8}	2.03×10^{8}	1.64×10^{8}
5	1.01×10^{8}	1.11×10^{8}	1.15×10^{8}	1.15×10^{8}	8.91×10^{7}
6	7.15×10^{7}	7.62×10^{7}	7.71×10^{7}	7.45×10^{7}	5.67×10^{7}
7	5.33×10^{7}	5.61×10^{7}	5.59×10^{7}	5.31×10^{7}	3.97×10^{7}
9	3.42×10^{7}	3.51×10^{7}	3.44×10^{7}	3.17×10^{7}	2.33×10^{7}
12	2.13×10^{7}	2.14×10^{7}	2.10×10^{7}	1.89×10^{7}	1.39×10^{7}
15	1.54×10^{7}	1.52×10^{7}	1.49×10^{7}	1.33×10^{7}	9.95×10^{6}
20	1.06×10^{7}	1.05×10^{7}	1.01×10^{7}	9.15×10^{6}	6.99×10^{6}
30	6.90×10^{6}	6.85×10^{6}	6.65×10^{6}	6.13×10^{6}	5.15×10^{6}
40	5.45×10^{6}	5.44×10^{6}	5.30×10^{6}	5.12×10^{6}	4.34×10^{6}
60	4.20×10^{6}	4.19×10^{6}	4.15×10^{6}	4.12×10^{6}	3.62×10^{6}
100	3.32×10^{6}	3.38×10^{6}	3.44×10^{6}	3.39×10^{6}	3.11×10^{6}
120	3.11×10^{6}	3.23×10^{6}	3.32×10^{6}	3.23×10^{6}	3.11×10^{6}

Source: Courtesy of André Maeder and colleagues in Geneva.

Stellar compositions are traditionally discussed in terms of abundances relative to iron, X/Fe (where X is the relevant element), as a function of the ratio of iron to hydrogen, Fe/H, this being the most convenient arrangement for observational purposes.

Stars with masses similar to the Sun are chosen in different populations of different ages, such as the thin or thick part of the disk, the halo or a globular cluster. Care is taken to ensure that their surfaces are not contaminated by internal nuclear processes, so that their compositions accurately reflect their date and place of birth. Red giants are therefore avoided. These stars

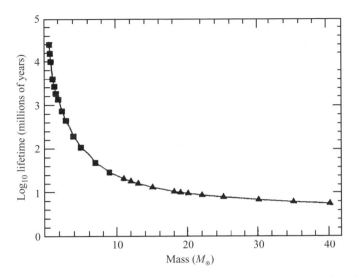

Fig. 8.1. Life expectancy of stars with solar metallicity. A star's lifetime depends principally on its mass at birth and varies little with initial metallicity. (From Riosi 2000.)

are subject to significant internal motions that carry the products of nucleosynthesis up to the surface. There are notable differences between the metallicities of the various subsystems making up the Galaxy, that is, the bulge, disk and halo.

The clearest indication of galactic evolution is the increase in iron content of the stars from one generation to the next. The youngest, those that formed in an older Galaxy, are richer in iron than their seniors, which formed when the Galaxy was younger. Hence, the iron deficiency of a star with respect to the Sun is all the more marked as it belongs to an older generation.

Everything suggests that the Fe/H ratio can be taken as a kind of chronometer, or at least as an index of evolution. It defines the chemical history of the Galaxy, and cannot decrease. The accumulation of iron in the interstellar medium is such that the abundance of this element increases monotonically, although in a way that is far from linear. The Fe/H ratio can be calibrated as a function of time by jointly determining the iron content and age of a great many stars selected from distinct generations. This then constitutes the basis of the age–metallicity relation.

Given the available data, galactic evolution can be divided into two eras, the halo era (metallicity between one ten-thousandth and one-tenth of solar values) and the disk era. The corresponding epochs are 1 and 10 billion years long, respectively. Concerning the history of the halo, two consecutive evolutionary regimes are discerned, with a transition at three thousandths of the solar metallicity.

Evolutionary beginnings are subject to whims and vagaries. In the dim and distant epoch when the halo formed, matter was not yet efficiently mixed up. The large variations in abundances observed at very low metallicities probably reflect the imperfectly homogeneous chemical composition of the gaseous medium produced by explosion of a small number of distinct and isolated supernovas, and subsequent sporadic contamination of the neighbouring interstellar medium. Stars of different masses would have exploded here and there, but they would not have supplied the same products. Their substance, poorly mixed into the interstellar medium, would have turned up locally in stars of the next generation, and these stars would thus carry the distinctive characteristics of the first exploding objects.

Explosions, or more precisely, the shock waves they induce, sweep through the interstellar medium and produce a cavity. The swept-up matter, rich in hydrogen, then accumulates to form a dense shell with ever-increasing radius. Inside the shell lies the hot, enriched matter ejected by the supernova, whilst outside is the ambient interstellar medium. We thus have a hot bubble surrounded by cold, dense gas. This is the structure of a supernova remnant. Mixing and interpenetration take place at the outer boundary. Tongues of hot matter penetrate the cold medium without, whilst the cold shell evaporates and dilutes gases within. When the shell reaches a diameter of 100 parsec and the mass of swept-up interstellar medium is around $50\,000\,M_\odot$, its speed has dropped to that of neighbouring clouds. It then breaks up and merges into the background. At the end of the day, the whole of the fresh delivery from the star will have been absorbed.

The changing abundance ratios observed at low metallicities (in the young Galaxy) are compared with the contribution from supernovas of different masses in order to determine which have enhanced the chemical content of the youthful Galaxy, and at what stage in its development. Measured abundances can then be used to adjust nucleosynthesis calculations, leading to more realistic supernova models. The genetic heritage of the most ancient stars in the galactic halo thus reveals their relationship with the first generation of supernovas (Fig. 8.2).

Organisation began to impose itself in the ensuing phases. Collective effects gradually wiped out the fluctuations of the early days, as the Galaxy changed almost imperceptibly from a set of individuals to a society. In other words, supernovas proliferated and their remnants started to overlap, until the interstellar medium began to take on a homogeneous composition. At this point, relative

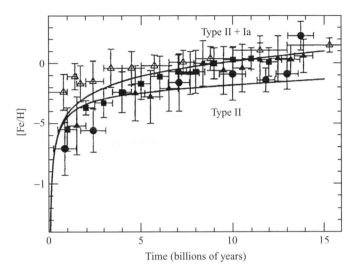

Fig. 8.2. Iron content as a function of the age of stars in the neighbourhood of the Solar System (age–metallicity relation). Age determinations are a delicate matter and somewhat uncertain. This explains the wide error bars and scatter of the data. Type Ia supernovas must be included to reproduce the observed iron evolution.

proportions of complex elements would have reflected the average contribution of a whole generation of stars. New stars forming in this well-blended medium would have had almost uniform composition.

Empirical correlations can be established between certain elements, in particular, those produced in the same categories of stars. These correlations can be brought out graphically by plotting the abundance of a certain element, say X, relative to another, say Y. When the data are plotted with X/H on one axis and Y/H on the other, using logarithmic scales for both, the slope of the curve generally lies between 0 and 2. A zero slope indicates a non-stellar origin. The best example is provided by lithium in halo stars, compared with iron (Fig. 8.3). This independence from metallicity indicates a primordial origin, that is, in the Big Bang. A slope equal to 1 indicates that SNII events are responsible for both the elements in question. This is the case for the α elements magnesium, silicon and calcium, but also for iron and beryllium.

A single type of object produces the various nuclear species in invariable amounts. A slope other than 1, or a change in slope, indicates the intervention of some other type of object.

If we restrict ourselves to the halo evolution, which we assume to be dominated by SNII events, given the very short time-scale involved, a statistical-type

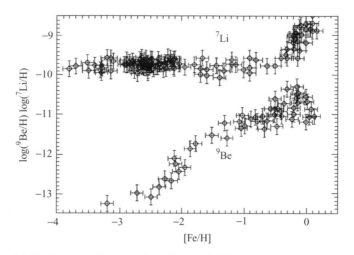

Fig. 8.3. Lithium, beryllium and iron. The symbol [Fe/H] denotes the logarithm of the ratio of Fe/H for the star and Fe/H for the Sun. The evolution of lithium and beryllium in the halo [Fe/H] < −1 is a classic example. The lithium content remains independent of the iron content in halo stars. This is known as the Spite plateau, named after the two French astronomers Monica and François Spite. It indicates a primordial origin (i.e. in the Big Bang). An upturn occurs just when the disk stars begin to take over. Beryllium is an archetypal example of elements created by spallation. Its abundance increases monotonically by accumulation as time goes by.

study of general trends leads to the following conclusions, which are most enlightening for supernova physics:

1. For several elements, differences come out from one star to another at the lowest metallicities (iron content). Not all stars with the same iron content, apart from a small number of exceptions, have the same abundance distribution, as already mentioned.
2. Lithium, on the other hand, displays almost no scatter and its abundance up to 0.1 $Z_\odot$ is independent of the iron abundance.
3. Beryllium and boron evolve in the same way as iron, showing that SNII events are responsible for these three elements.
4. Relative to iron, the α elements ^{24}Mg, ^{28}Si, ^{32}S, ^{36}Ar and ^{40}Ca are uniformly more abundant than in the Solar System, in agreement with nucleosynthesis calculations (Fig. 8.4). We deduce that the galactic halo is a good place to study nucleosynthesis due to gravitational-collapse supernovas, for the contents of stars in the halo would appear to have been composed by these massive stars with their consequently short lifetimes. It was long thought that oxygen evolved in a similar way, but recent data have

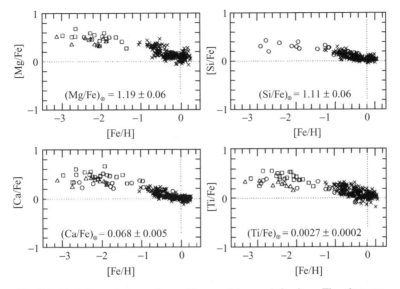

Fig. 8.4. Evolution of magnesium, silicon, calcium and titanium. The elements Mg, Si and Ca are known globally as α elements. They occur in excess relative to iron in halo stars. It is surprising to find Ti on the list, given that it belongs to the iron group. (From Arnould & Takahashi 1999.)

sown the seed of doubt over its behaviour at low metallicities. According to some research workers, it would seem that oxygen behaves in a different way to its α siblings and that the ratio O/Fe, rather than remaining constant, actually grows at low metallicities. However, this trend has not been unanimously confirmed.

5. Titanium behaves like the α elements, whereas one would expect it to resemble a ferrous metal.

6. Discrimination grows between odd and even at low metallicities, at least for the light metals, such as sodium and aluminium. The even–odd imbalance is much less significant for the ferrous metals. Thus, chromium and manganese manifest a comparable underabundance for metallicities below 0.003 $Z_\odot$. In contrast, cobalt, which is odd like manganese, is overabundant in the same metallicity range.

7. The heavy elements show varying trends but, on the whole, strontium follows iron, whilst barium is underabundant below 0.01 $Z_\odot$. It is observed that, in every case, marked differences occur between stars with the same iron content.

8. The abundance distributions of certain stars investigated in detail conform to that of the r process.

A tedious lesson

Armed with these suggestive prolegomena, we may now turn to the key problem of chemical (isotopic) evolution in galaxies, starting with our own.

The stars are the main driving force for nuclear evolution. However, they are not all equally productive. Let us therefore distinguish between three categories of object, depending on their masses: small, intermediate and large. The first, less massive than the Sun, lead to a finished product which is just what we started with. They are more or less sterile, evolving so lazily that, after 10 billion years, they have not yet delivered up a single nucleus. Along with compact stellar corpses, these are a genuine dead weight to galactic progress. It is as though their contents had been locked up forever in a bank safe.

The second group, with masses between one and eight times the solar mass, experience an agitated evolution, tacking back and forth across the HR diagram, so that it is difficult to keep track of them. At the end of their evolution, they move into a planetary nebula phase, eventually leaving a white dwarf for posterity.

However, the work awards go, without possibility of dissension, to stars in the third group. With masses between 8 and 100 times the solar mass, these are the stars that will one day explode. Their self-sacrifice earns them special attention and this is indeed what we have been attempting throughout, with all due respect.

Apart from their productivity, another important feature is the lifetime of these stars as a function of their mass, since we are talking about the gradual metal enhancement of the Galaxy. These lifetimes should be compared with the age of the Galaxy, of the order of 10 billion years. In fact, if these stars are eventually to make a donation to the general well-being of future generations, it will often happen at the end of their own existence. Low-mass stars shine economically and only open their box of treasures at the end of a correspondingly long life. In contrast, heady and extravagant, high-mass stars exhaust themselves in maintaining their brilliance and scatter their vast opus in a dazzling explosion. As far as type Ia supernovas are concerned, their offering is much delayed by the fact that one of the two protagonists must first transform itself into a white dwarf and that the other must then administer the lethal dose of stellar material, all of which takes precious time.

By good fortune, the effects of metallicity on stellar lifetimes are rather limited, being only of the order of a few percent. In a first analysis, it is enough to apply the mass–lifetime relationship calculated for stars with solar metallicity.

To sum up, high-mass stars are the main suppliers of complex isotopes. All elements from carbon to calcium are synthesised inside them by the relatively gentle and slow process of hydrostatic combustion, whilst iron and its kin, together with the r-process isotopes, are generated in the final explosion (SNII event). Exploding white dwarfs (SNIa events) add a decidedly ferrous touch to the result. Most of the carbon, nitrogen and minor isotopes of oxygen originate in intermediate-mass stars (2–8 $M_\odot$), as do the s-process isotopes.

Having reviewed the courageous stellar workforce, assessing each for profitability and picking out the massive stars for their exemplary attitude, we now believe ourselves capable of setting up a genuine stellar sociology. The aim is to extend the theme of nuclear evolution from stars as individuals to the Galaxy as a society.

Picking up from the Big Bang, the story of nucleosynthesis is mainly tied up with the physics of stellar evolution and its nucleosynthesis, and by environmental factors, such as the existence of cold, dense clouds, propitious for star formation. There is no hope of understanding such processes by theory alone. The best way to comprehend the history of the chemical elements in our Galaxy, one after the other, is to look for fossils. These can be found in low-mass stars with very long lifetimes, comparable with the age of the Galaxy itself. In their outer envelopes we expect to find preserved their composition at birth. Born from the debris of the first massive stars, they are still shining in the halo and their light carries the genetic characteristics of their precursors. These stars are invaluable for the chemical genealogist, for they are relics from the ancient history of lithium, carbon, oxygen, iron, yttrium, europium, and so many others.

The study of galactic evolution thus comprises two aspects. One is of an observational nature, wherein abundances are assessed in stars of various generations, by means of spectral analysis. The other is purely theoretical, involving numerical simulation of the chemical evolution of the interstellar medium in which stars are born.

The first step requires highly accurate and careful observation, but also finely tuned models based on atomic physics, to extract the relative abundances of elements from stellar spectra. Even the rarest of the elements are thus revealed. The second step requires an overall view of the cycle of matter in the Galaxy, whereby intermingling clouds and stars manage their flock of atoms, identifying the factors which determine this organisation. The complexity of the problem is such that we may only describe the main lines. The authors of this great portrayal of the origin and evolution of the elements must clearly recognise the limitations of the exercise. This assumes a certain realism on their part. The problem is to harmonise an idealised model with data that are often incomplete, inaccurate and even ambiguous, in such a way as to write the universal saga of the elements, assumed to repeat itself in every galaxy. The cosmic scriptwriters propose various versions of the same tale, hoping that each will be more realistic

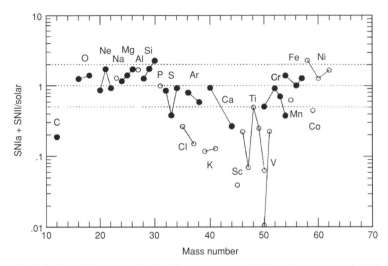

Fig. 8.5. The delicious cocktail of the supernovas. Mixing 13 measures of SNII with 1 measure of SNI, we find a composition of matter that approaches observed abundances in the Solar System. Certain isotopes of chlorine, potassium and scandium, among others, are not produced in sufficient quantities, however. (From Nomoto *et al.* 1997.)

than the others. But the conceptual framework is the same. One might say that the profusion of grapeshot makes up for a certain inaccuracy in the aim.

Results

Arriving at the solar abundances

The first criterion for success of a theory of nucleosynthesis and galactic evolution is obviously to explain the measured abundance distribution in our own galactic neighbourhood. In other words, we compare with the abundance table discussed at the beginning of the book. It is gratifying to observe that, for the most abundant nuclei between hydrogen and zinc, this comparison gives good results.

SNII events alone explain the observed solar abundance distribution between oxygen and chromium. This can be taken as a major theoretical achievement. Complementary sources of hydrogen, helium, lithium, beryllium, boron, carbon and nitrogen are required, and these have been identified. They are the Big Bang, cosmic rays and intermediate-mass stars. Around iron and a little beyond, we must invoke a contribution from type Ia supernovas (Fig. 8.5). These must be included to reproduce the evolution of iron abundances, a fact which suggests that about half the iron in the Sun comes from this type of object and the other half from SNII events.

Abundances of α elements relative to iron

The α/iron abundance ratios fix time-scales for processes involved in the compositional evolution of the Galaxy. For example, SNIa events resulting from the force-feeding of white dwarfs are big iron producers. It turns out that, even when pushed beyond the limit, they have longer lifetimes than SNII precursor stars, which are the main sources of oxygen, silicon, sulphur and calcium. The lapse of time separating hasty SNIa events from the first SNIIs (at least 100 million years) explains the high value of the O/Fe ratio observed in halo stars. The change in slope of the O/Fe evolution curve marks the arrival of the very first SNIa events on the stage.

Even–odd effects

Even nuclei, and in particular the class of α nuclei (oxygen, magnesium, silicon, calcium), are the basic products of nucleosynthesis in high-mass stars. They are abundantly present in the ashes of SNII events, where the α/iron ratio is about three times the solar value. The amounts of even elements ejected by explosion of a high-mass star are, to the first approximation, independent of the star's initial metallicity.

The same cannot be said for the odd elements nitrogen, fluorine, sodium, aluminium, phosphorus, chlorine, potassium, vanadium, manganese and cobalt. The quantities produced depend significantly on the initial composition in the sense that a star with low metallicity is likely to produce fewer odd elements than a star with high metallicity. At the root of this tendency lies the fact that the major part of the metals (CNO) available to the star at birth are transformed into ^{14}N, then ^{22}Ne, via the chain reaction

$$\text{CNO} \longrightarrow {}^{14}\text{N},$$
$$^{14}\text{N} + \alpha \longrightarrow {}^{18}\text{F} + \gamma,$$
$$^{18}\text{F} \longrightarrow {}^{18}\text{O} + e^+ + \nu \quad \text{(weak interaction)},$$
$$^{18}\text{O} + \alpha \longrightarrow {}^{22}\text{Ne} + \gamma.$$

Consequently, the n/p ratio in the star increases in proportion to the initial CNO content. This happens thanks to the weak interaction, which systematically transforms protons into neutrons when the latter become too abundant, resetting the balance between N and Z in favour of nuclear stability. The weak interaction therefore constitutes a kind of internal regulator, favouring equilibrium by constantly bringing deviant nuclei back into the valley of stability.

It should be noted that odd elements produced in explosive nucleosynthesis depend less on metallicity than their counterparts fashioned by slow (hydrostatic) nucleosynthesis, for the n/p ratio is steadily modified by various weak interactions operating in the advanced stages. In other words, the n/p ratio deep down in the star, in regions affected by explosive nucleosynthesis, no longer reflects the initial ratio inherited from the interstellar medium. At least, this is what calculations suggest. However, the cause of all these phenomena remains relatively obscure, given the complex way in which nuclear reactions are interwoven within massive stars in the advanced stages of their evolution.

In fact the weak interaction becomes much more effective once core oxygen combustion gets under way. The n/p ratio in the core of massive stars is modified under the effect of electron captures

$$p + e^- \longrightarrow n + \nu \,.$$

Neutronisation has begun. The n/p ratio in the nuclear potpourri is one of the factors influencing nucleosynthesis. A mixture well provided for in neutrons favours the construction of species rich in neutrons (see Appendix 3).

Manganese is an odd element that is underrepresented in halo stars. However, the above explanation does not seem to apply since scandium ($Z = 21$) and vanadium ($Z = 23$) do not follow the same trend, no more than does cobalt ($Z = 27$). It is tempting to deduce from the unusual behaviour of manganese that it is produced by SNIa events and that we are observing a mirror-image phenomenon to the one described for α elements.

Heavy elements

Rather unexpectedly, a striking similarity is observed between proportions of r-process elements measured in a handful of ancient stars and the Sun. The hallmark of the r process thus appears very early on, indicating the operation of a rapid and efficient process in the very first stages of galactic evolution.

We conclude that the process responsible for the production of r-type isotopes beyond barium has operated in the same way since the origin of the Galaxy. The process is therefore unique. There is little risk of error in suggesting that it is related to SNII events. We may then define the conditions of neutron irradiation and thermodynamic parameters relevant to this nucleosynthesis, without necessarily being able to establish the detailed mechanism.

Promising scenarios appealing to a high-entropy neutrino wind were first outlined, but quickly discredited. There is no escaping the fact that today we

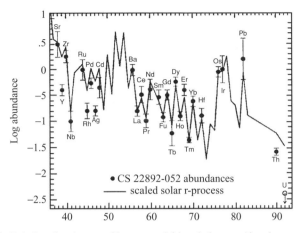

Fig. 8.6. Relative abundances of heavy nuclei in a halo star. Abundances are nor-malised to the barium abundance. The continuous line represents r abundances (from Käppeler). The excellent agreement suggests that previous nucleosynthesis was dominated by the r process and that the star CS 22892-052 formed from the debris of a type II supernova. (From Sneden 2001.)

still cannot put forward a specific site for the r process which gives global agreement with measured abundances. It is also conceivable that there might be not just one r process, but two, operating in high-mass stars, but nevertheless in different mass ranges. In this respect, the current situation in astrophysics is thus unstable, even explosive!

The s process is slow to start moving, for it is related to stars in the asymptotic giant branch. These have a maximum mass of 8 $M_\odot$, implying a lifetime of at least 20 million years. It is not surprising then to observe that abundances in old halo stars carry a clear r-process signature (Fig. 8.6).

The detection of thorium in stars of very low metallicity by Patrick François and Monique and François Spite has opened the way to a direct determination of the age of the most ancient stars, and hence also of the age of the Galaxy. The age of some of the oldest halo stars has been estimated at 15.6 billion years, with an error margin of 2 billion years. This agrees with the classic determination of the age of the oldest globular clusters at 14.9 ± 1.5 billion years, and the age of the Universe by means of remote type Ia supernovas at 14.2 ± 1.7 billion years (Arnould & Takahashi 1999).

To sum up the results of a long series of measurements, abundances of heavy elements in the halo are characterised by a significant r component and a con-siderable variation in abundance ratios at very low metallicities. The typically r-type abundances combined with the wide scatter in these abundances in ancient halo stars provides evidence in favour of type II supernovas. The gradual reduction in scatter at higher metallicities attests to an ever-greater homogeneity of the interstellar gas as the number of supernovas increases.

On the whole, the predictions of nucleosynthesis are thus borne out by observation, an undeniable success for the theory.

1. The evolution of the α elements is well understood (except for the latest oxygen measurements in halo stars, which are still uncertain), assuming that iron is produced by type Ia supernovas in the galactic disk as well as by type II supernovas.

   The first SNII events only enrich the halo, given their precocity. Their signature is a high α/Fe ratio. These are followed by the disk phase, where the SNIa come on the scene. The distinctive feature in this case is a very low α/Fe ratio. Adding the two produces a gradual reduction in the ratio. Thermonuclear supernovas (SNIa) are latecomers, arriving when all star formation has ended in the galactic halo and the disk takes over.

2. The light metals sodium, magnesium and aluminium are adjacent on the periodic table and have a common origin. They are found in profusion in the ashes from gentle, non-explosive combustion of carbon and neon. Production of sodium and aluminium grows as the Galaxy evolves.

3. Heavy nucleus abundances in ancient stars are determined by rapid neutron capture, very probably associated with type II supernovas.

Some grey areas remain, along with certain thorny problems. These are observational, as in the case of oxygen abundance measurements in halo stars, but also theoretical, concerning convection, or the cutting of the Gordian knot between neutron star and ejecta. They have deleterious effects when calculating the amount of iron produced by supernovas of various masses. Fortunately, this inability to calculate the mass of ejected iron, that is, to make the Gordian cut, is balanced by the fact that we can measure this quantity directly by analysing supernova light curves. In order to explain the strange Cr/Fe, Mn/Fe and Co/Fe ratios in the poorest halo stars (with iron contents between 10 000 and 300 times lower than the Sun), it has been suggested that the position of the cut depends on the mass of the progenitor star. Data indicating a manganese and chromium deficiency and a cobalt excess relative to iron in ancient stars can be explained by shifting the boundary between falling and ejected matter towards the interior of the star. It seems that this explanation agrees with the amount of nickel-56 ejected, as determined from the light curves of individual supernovas. This rather delicate empirical fact calls for a detailed physical explanation, but its discovery should help in perfecting explosion models.

However, this is not the end of the story. Calculations carried out in a spherically symmetric context must now be extended to include clearly or subtly anisotropic effects, in order to model jets, rotation and the like. Three-dimensional numerical simulation is required. Astrophysicists should profit from considerable progress made in hydrodynamics with the development of extremely powerful lasers in France and the United States, but that is another story.[1]

These are therefore the main lines of research into galactic evolution. However, stellar archaeology is still in its early stages. In future years it will obtain a significant boost from the VLT and its high-quality spectrographic observations, to become eventually a major branch of astronomy.

Scientific hopes and aspirations

One cannot overemphasise the importance of accurate abundance measurements and estimates of the errors involved in them. If suitable data are unavailable, or if their quality cannot be properly assessed, it is quite impossible to draw astrophysical conclusions. In this respect, systematic observation campaigns are preferable, collecting large and uniform samples of good quality. One might seek to achieve the aims on the following list:

- to carry out a detailed study of the evolution of carbon and also nitrogen, which up to now has defied all attempts at understanding;
- to establish a reliable relationship between oxygen and iron at low metallicities in order to determine the production mechanism for light elements in the halo;
- to make a careful study of the intermediate metals sodium, magnesium and aluminium in order to test the theoretically predicted even–odd effect;
- to carry out a precise study of the iron peak, examining chromium, manganese and cobalt abundances in the low metallicity regime, so as to elucidate their origins and refine supernova models, thereby clarifying the position of the Gordian cut;
- to make joint measurements of europium and barium as a function of metallicity, as well as a whole range of heavy elements, in order to map out the gradual rise of the s process in the galactic halo.

---

[1] Recognition is due to Jean-Pierre Chièze and his teams at the CEA in France. The detailed study of binary star systems with mass transfer leading to an explosive situation should be carried out in parallel, as should studies of the explosion itself. Here again, the megajoule laser will be a great boon to research.

Lithium, beryllium, boron and the controversy over oxygen and the origin of the light elements

As the halo stars tell us, oxygen was overabundant with respect to carbon and iron in the adolescent Galaxy. This has been well established for a number of years now. However, debate has rekindled over the amplitude of this oxygen excess and its variations as a function of metallicity.

A large increase in the O/Fe ratio in stars at low metallicity was reported by Israelian *et al.* in 1998 and by Boesgaard *et al.* in 1999, contradicting earlier data which suggested an approximately constant O/Fe ratio. Now oxygen is particularly relevant to the astrophysics of cosmic rays. This is because spallation products under collision include the light nuclei lithium, beryllium and boron.

Galactic cosmic rays are essentially made up of fast-moving protons (and helium nuclei). We may therefore assume that protons fragment a small fraction of the oxygen nuclei floating in the interstellar medium, reducing them to lithium, beryllium and boron. We will call this the pO mechanism. We may also imagine that oxygen nuclei ejected directly and accelerated by supernovas could shatter upon impact with the many hydrogen nuclei in the interstellar medium, to produce once again the precious lithium, beryllium and boron nuclei. We call this the Op mechanism. Since these two mechanisms are both feasible, we need to determine their relative contributions at each stage in the development of the Galaxy.

It is known that the oxygen abundance in the interstellar medium increases all the time: this nucleus is produced by type II supernovas which, one after the other, also contribute their iron production to the Galaxy (Fig. 8.7). The pO mechanism is thus likely to grow in importance as the Galaxy evolves. In other words, clues to the Op mechanism should be sought in the early phases of galactic evolution, that is, in halo stars. The fact remains that the two mechanisms induce different evolution in beryllium and boron as a function of oxygen.

The Op mechanism leads to proportionality between oxygen and beryllium abundances, for example, because these two elements arise from the same source, namely, type II supernovas, oxygen directly and beryllium indirectly (via collisional disintegration of oxygen into beryllium). A constant Be/O ratio, independent of O/H, would be the hallmark of the Op mechanism.

Proportionality between Be/H and the square of O/H would correspond to a pure pO process.[2] Any intermediate situation would indicate a mixture of the two processes. The main step to take here is to establish an accurate and

---

[2] This quadratic relationship results from the fact that the number of oxygen atoms is proportional to the cumulative number N of supernovas and that the number of high-energy

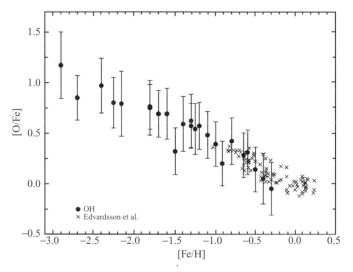

Fig. 8.7. Evolution of oxygen. Unlike the other α elements, oxygen continues to grow more abundant at low metallicities. This effect remains unexplained. However, the validity of the data has been contested. (Courtesy of Boesgaard and co-workers.)

calibrated Be–O diagram. However, for practical reasons, actual measurements concern Fe rather than O in the relevant stars. The conversion of Fe to O is therefore the key problem.

If Fe and O are proportional, as was thought until very recently, we conclude that the Op process dominates in the halo. If on the other hand there is an excess of oxygen relative to iron which becomes more marked at low metallicities, the pO process is amplified and favoured.

The oxygen abundance in the interstellar medium and its evolution in the early Galaxy depend on the adopted relationship between oxygen and iron, and with it the quantity of heavy elements produced when oxygen is shattered by galactic cosmic rays. Observers still disagree on this key issue. It has become urgent to find some common ground because our explanation of the origin of the light elements in the early Galaxy depends upon their verdict.

For the moment, the situation remains confused but it should soon be clarified by precise measurements of oxygen and iron in stars with very low metallicities, the kind of measurements now within reach of the VLT.

protons is proportional to the rate of change dN of that number with respect to time. The abundance of spallation products, that is, the integral of the product $N dN$, is proportional to N^2 and hence to the square of the oxygen abundance. The formal details are given in the article by E. Vangioni-Flam *et al.* (2000) *Physics Reports* **333**, 365.

Depending on whether it is the new or the old preference that wins out, that is, depending on the magnitude of the oxygen excess with respect to iron, light-element production will be imputed to the good old galactic cosmic rays, which are nothing other than accelerated interstellar matter, or else to a distinct component of these rays emanating directly from type II supernovas. The latter are grouped together in superbubbles which they hollow out in the interstellar medium. One supernova injects fresh nucleosynthesis products, including oxygen, whilst the following accelerates a certain fraction of the available oxygen nuclei up to high energies via the shock wave it generates. From this point on, these nuclei are likely to shatter on the first collision amidst the throng of hydrogen and helium nuclei that populates the interstellar medium.

The gas and dust of the interstellar medium

It is very useful to complement the compositional analysis of stars by a like analysis of the interstellar medium. This can be done by making use of absorption lines which the latter removes from the UV spectrum of hot, bright stars (Fig. 8.8). Measured abundances only concern gases lying between the source star and the observer. Matter contained in dust grains escapes detection.

Comparing interstellar and solar abundances, it is found that there is a significant lack of elements with an affinity for the solid state (refractory elements), such as iron and nickel. Condensation into the solid state and the concomitant impoverishment of the gas affect different elements to varying degrees. For example, oxygen and zinc, like the rare gases, are practically immune to this effect. The fact that zinc conserves its gaseous state provides an invaluable metallicity indicator when studying remote extragalactic clouds, which are of the same nature as interstellar clouds. The key point here is that depletion in the gaseous phase affects different elements to varying degrees depending on how volatile or refractory they happen to be.

Cosmological clouds

Needless to say, nuclear astrophysics does not limit itself to the chemical evolution of the Milky Way, but seeks to bestow a cosmic dimension upon its quest for the origin and evolution of the elements. Ever beclouded by the earliest times, the time of the genesis, the astrophysical spirit infiltrates the turbulent youth of the galaxies, going back to the stormy days of their early childhood.

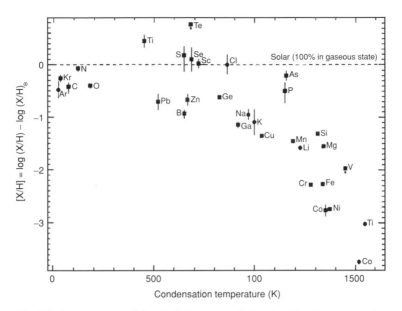

Fig. 8.8. Imprisonment of chemical elements in dust grains. The elements precip-
itate out to varying degrees to form grains, depending on their affinity for the solid
state. Volatile elements, with low condensation temperature, stay for the main part
in the gaseous state. Refractory elements, with high condensation temperature,
are mainly imprisoned within dust grains. Only atoms in the gaseous phase are
detected by classic techniques analysing UV absorption spectra. The light source
whose spectrum has been decoded here is the hot star ζ Ophiuchi.

But as we move back in time, the mists gather. Lacking stars, astrophysicists
must satisfy their curiosity with clouds alone.

The extremely luminous objects known as quasars are used here as beacons.
Their light is intercepted by huge numbers of atoms distributed in clouds and
astronomers can repeat their interstellar performance on a cosmic scale. The
absorbing clouds are now intergalactic rather than interstellar, but the technique
is the same. Absorption lines are picked out and the composition of the absorbing
medium is determined by analysing their position and depth. The speed(s) of
the cloud(s) and their distance(s) are deduced from the spectral shifts of their
lines, according to the well known cosmological procedure.

Analysis is carried out on certain types of clouds, called damped Lyman
alpha systems (or DLA systems). These clouds are essentially neutral, with
high column densities, of the order of 10^{20} atoms per square centimetre.

The interest of this method is to make accurate abundance determinations
at distances (epochs) that would be inaccessible to conventional stellar and
galactic astronomy. The only problem is that the absorbing systems belong to

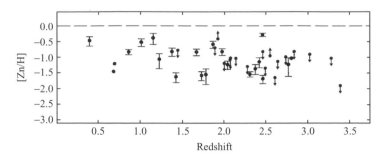

Fig. 8.9. Zinc in cosmological clouds.

galaxies of unknown type, because the underlying structures cannot be made out. Indeed, this is attested by the disparate values obtained for abundances at a given redshift.

Along the line of sight towards a quasar lie huge quantities of absorbing material, with varied composition and structure. Measured abundances concern only the gaseous component of this material. For this reason, abundances are distorted by dust, which subtracts atoms in varying amounts depending on their affinity for the solid state.

However, zinc escapes incarceration in the dusts of our own Galaxy, and there is every reason to think that it will do the same in absorbing clouds along the very much longer line of sight to quasars. At least, this is the hypothesis that is usually made. On the other hand, iron is heavily afflicted by the precipitation effect. It cannot therefore be used to trace back the evolution of the galaxies and has to admit the superiority of zinc in this respect. This transfer of power is all the more eagerly accepted in that the Zn/Fe ratio is constant down to very low metallicities in halo stars of our own Galaxy.

This observation nevertheless contradicts the relevant models, which suggest that the amount of iron produced by supernovas depends sensitively on metallicity. This in turn indicates that the nucleosynthesis of zinc is far from being well understood. But once again, the facts take precedence.

Zinc has thus been promoted to the rank of prime evolutionary indicator for galaxies or protogalaxies associated with DLA absorbers. With its help, we aim to measure the chemical evolution of the Universe, because it leaves its signature on the spectra of the most distant objects and because it is predisposed to the gaseous state (Fig. 8.9).

The evolution of metallicity as a function of redshift, and hence as a function of time, provided that we have selected our cosmological model, teaches us mainly about the evolving rate of SNII explosions, for it is in these that zinc originates. It thereby informs us of the rate at which massive stars were forming, and consequently the rate at which all stars were forming, if we can assume an invariable and universal mass distribution (see Appendix 5).

The evolution of abundance ratios X/Zn as a function of redshift gives us an idea of the relative rates of SNIa and SNII events and how they evolve.

Relative abundances from DLA absorbers can be used to trace back the evolution of the galaxies (or galactic haloes) that contain them over considerable spans of distance and time. One would like to compare cosmic evolution (Zn) to galactic evolution (Fe) over the same period of time. One would also like to observe in DLA systems the same overabundance of α elements noted in galactic halo stars. This would strengthen the idea that all galaxies are significantly impregnated with SNII events.

It is easy to understand the lively interest in these clouds. However, for the time being, results have not lived up to expectations.

It is surprising to note that zinc in DLA systems hardly evolves at all over a range of redshifts from 4 to 0.5, whereas star formation, deduced from the changing colour of the galaxies as a function of z, would indicate the opposite effect.

This raises a doubt over the representative nature of DLA systems, and even suggests that there may be some selection effect in the observations. As Patrick Boissé and co-workers have demonstrated, DLA systems may contain large quantities of dust which would obscure their gaseous content and make them difficult to detect.

Data show a considerable statistical scatter, much greater than the error bars attached to individual measurements. This suggests that a wide variety of galaxies of every morphological type (elliptical, spiral and irregular) is involved here, and that lines of sight may intercept a number of incongruous objects. In other words, DLA absorbers may constitute a whole zoo of different systems. Indeed, the solution to the problem may lie in a type of analysis inspired by zoology, namely, to improve the statistics of the sample, identify categories on the basis of common properties and subdivide populations. Maybe then more intelligible characteristics will be drawn out. Once again our thoughts turn to the VLT.

9

Conclusion

Modern cosmology is a physical and mathematical tale, telling of the creation of the Universe from nothing, or almost nothing, and describing its composition, structure and evolution. All the subtlety of this story lies in the word 'almost'. A perfectly rational discourse on the origin of the Universe, taken in the absolute sense, would nevertheless appear to be impossible because in the beginning the terms 'time', 'space' and 'energy' are undefined. Zero time is an instant in time that does not yet exist. The quest for the origin, or rather some mirage of the origin, remains the principal driving force in cosmology. However, in order to escape from the self-contradictions of the initiating event and conceptual catastrophes it triggers, one might assign a less ambitious aim to astronomy and its related sciences. For example, one might begin simply by trying to give meaning to the words 'Universe', 'matter', 'light', 'Big Bang', 'star', and now 'quintessence'. Spelling out this cosmic semantics, we arrive at the following (provisional) definitions.

The Universe is what extends, proving itself through an expansive motion, which seems to accelerate without respite and without hope of return, and through its evolution, an irreversible advance towards more complex atomic structures. Its components are matter and quintessence. Matter is what has weight, gravitating and curving space. In this sense, light is matter in a neutral material form. Quintessence is a latent state of nature, being invisible and impalpable. Modern physics attributes to it an energy content and a repulsive gravitational effect.

Light and matter are subject to mutual transformation. The Big Bang is the event which transformed light into matter. Stars are places where matter is transformed into light. These transformations are only partial.

Heat is reconstituted in each star. Condensing and heating up the matter within them, stars are the antithesis of the Big Bang, making up for the nuclear shortcomings of the initial event, due to an over-lively dilution, by a slow concentration. They manufacture carbon and the higher elements. In the general economy of the Universe, they thereby play the role of atom suppliers and life providers, for the heavy elements, combined with primordial hydrogen, link themselves up into molecules in the cool shade of huge interstellar clouds. A tiny fraction of atomic matter goes to make human beings. Indeed, the human species carries little weight before the mighty Galaxy: the ratio of their masses is of the order of 10^{-31}. Could it be this butterfly lightness that sends humankind rushing towards the flames of these beginnings?

The geometric history of the Universe (its expansion) imposes a series of metamorphoses on matter and radiation, but not on the quintessence, which remains essentially constant. It escapes dispersion and ends up dominating over other energy forms.

How is life possible in a Universe under constant threat of dilution and cooling by expansion and the quintessence that feeds it? For this, we must thank gravity. Although the mean density and temperature of the Universe can only decrease as time goes by, under the effect of the expansion of space, the local attraction of matter by matter has accomplished its constructive work of isolating certain structures and sheltering them from universal divergence, giving birth and prosperity to stars in the gaseous swirl of the galaxies.

From a thermal standpoint, cosmic history can be summed up as a generalised cooling trend induced by the dilatation of space. However, here and there, in certain regions, stars concentrate and bear matter to high temperatures, as balls of incandescent material floating in a cold ocean.

In this respect the Milky Way is like a nature reserve, housing a complete flora of stars, from whose petals the good seed has flown. The flesh of future humanities is there, in the debris of exploded stars. In the beginning, the Galaxy was gas without stars. In the end, it will be stars without gas. It is gently fading as gas supplies dwindle. As it does so, the element with mass number 56, iron, the hard-hearted master promised such a fine future by its noble constitution, will never surpass the ancient helium. The most perfect is not the one that reigns, at least, once we go past number 1, hydrogen.

When generation after generation of stars had gone by, a modest body separated out from its parent cloud at the Galaxy's edge and called a retinue of planets from the disk around it. On one of these, life emerged, then consciousness. Today, thinking matter is looking back upon and trying to understand its inert past, so stellar and nebulous. One day, the Sun will die, huge and red, victim of its own excesses, and all the atoms of the dead, and of the stones and flowers, will return to it. The atoms of Earth will be restored to the sky.

The existence of atoms is now fully recognised. We have been able to determine their cosmic sources, the culmination of an admirable theoretical and experimental effort by atomic and nuclear physicists. But it is in its moment of triumph that the atom slips from our grasp and takes its last bow. Suddenly, the sky has been undone of its atoms and nuclei. For this is the age that cries out the cosmic insignificance of the atom, just as it was enthroned. Today, a great many cosmological studies suggest that about two-thirds of the Universe exists in the form of a negative pressure component, the quintessence, exerting a repulsive form of gravitation, whilst the other third is composed of non-atomic dark matter. In the future, statistical studies of high redshift gravitational lenses, combined with a better understanding of the cosmic background radiation and the large-scale distribution of matter, should make it possible to determine the physical characteristics of the invisible energy hidden throughout space.

As the concept of the Universe has extended, astronomy has experienced a series of Copernican revolutions. In the beginning, the Universe was just the Solar System, as propounded by Aristotle and Ptolemy. It was then identified with the Galaxy, and finally, with the collection of all galaxies. Several millennia of astronomy have taken us to the following understanding:

- the Earth is not at the centre of the Solar System;
- the Sun is not at the centre of the Galaxy;
- the Galaxy is not at the centre of the Universe;
- the star is the mother of the atom;
- atomic matter is a mere froth upon the main matter of the Universe.

The Universe has no centre. There is no privileged place to mark it out. Has the bell tolled for an anthropocentric world view? This is not my own view (Cassé 1999). Space is lost, but time has been reinstated.

We belong to the great age when matter speaks. Humankind has invented science to heal cosmic amnesia.

The conspiracy between Earth and sky is much more closely engineered than the astrologers ever imagined. Their geometrical link has been replaced by a genealogical one. The star is the mother of the atom. The sky that held our destiny is no longer. We are the orphans of the sky, for we have lost our zodiac. But we have found the Universe.

There, where order seemed to reign in the mechanical silence of death, a life-giving exuberance has sprung. The path that leads from the multitude of anonymous and abstract particles engendered by the Big Bang to the infinite variety of forms and states, to the moving intimacy of the things we know and love, this path has to go through the star, whilst the star must come from the cloud, the cloud from light, and light from the primordial void. This is the physical chain of the genesis, a genealogy of matter.

Such was once the good fortune of the atom. There were secret bonds and nuclear wedlock, and behold! The whole history of time is thrown open for all to see. Lucretius sings once more and supernovas burst out laughing in the face of the Universe.

Appendix 1

Invisible matter and energy

Glossary

baryonic dark matter dark matter made up of baryons that are not luminous enough to be observed

Big Bang initiating event in cosmology

boson particle with integral spin

critical density dividing line between an open and a closed universe

dark matter unseen and possibly invisible matter

fermion particle with half-integral spin

gravitational lens dense and massive object capable of deflecting light from more remote celestial objects, thereby multiplying and intensifying their images

neutralino hypothetical particle predicted by so-called supersymmetric particle theories

non-baryonic dark matter dark matter made up of neutrinos or neutralinos

quintessence substrate exerting a gravitational repulsion

rotation curve radial variation in the orbital speed of stars and gases around the centre of a spiral galaxy

supersymmetry an extension of elementary particle theories which allows for counterparts to the usual bosons and fermions

Does invisible matter exist?

Two arguments support the idea that some invisible substance exists in the Universe. The first is dynamic. It starts from observation of motions under the effect of gravity. The second is related to Big Bang nucleosynthesis, i.e. nuclear cosmology, which combines cosmology and nuclear physics.

Dynamical proof

The first observation is that, if we can go by what light is telling us, most of the matter in the Galaxy (and indeed any galaxy) is concentrated in the galactic bulge, a marked, reddened swelling at the centre of the star distribution. Therefore, if we assume that the mass distribution of luminous objects is representative of the total mass distribution in galaxies, every spiral galaxy should behave like a vast Solar System, with the stars and clouds playing the role of planets and the galactic bulge that of the Sun.

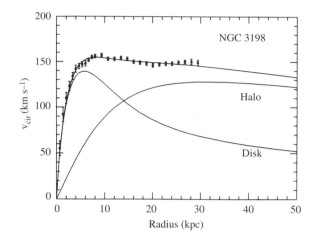

Fig. A1.1. Orbital speed as a function of distance from the centre of the galaxy NGC 3198. The flat part of the curve can only be explained under the assumption that there is a massive halo of dark matter. Points correspond to observations. Curves show contributions from the disk and the halo, calculated using a suitable model.

According to Newton's law, the circular speed of an object in orbit around some central point should decrease with distance from that point. However, in the case of disk-shaped galaxies such as our own, it is observed that the circular speeds of component stars and gases are roughly independent of the orbital radius beyond a certain distance from the centre. Stars in the outer regions of galaxies revolve too quickly around the common centre to uphold the edicts of the laws of gravity, unless some unseen matter is present (Fig. A1.1).

Let r be the orbital radius, m the mass of a test object, and $M(r)$ the mass of that part of the galaxy lying within radius r. The equation describing equilibrium between gravitational attraction and centrifugal force is

$$\frac{GmM}{r^2} = \frac{mv^2}{r},$$

where M is treated as a function of r. We can account for the phenomenon described above by setting $v(r)$ constant in the above equation. It then follows that M must be proportional to r. This implies that the mass is not confined to the galactic bulge, but spread out in a uniform manner through a sphere extending much further than the luminous region of the galactic disk. An enormous halo of invisible matter must therefore encompass the Milky Way and most other spiral galaxies.

It is thus assumed that the (rotational) speeding offences committed in the galactic periphery are due to the existence of a massive halo of invisible matter. In our own Galaxy, there must be ten times as much dark matter as visible matter, amounting to some 1000 billion solar masses. We may deduce that the same is true of all spiral galaxies, that is, they possess haloes about ten times as massive as their visible disks.

In a moment of dark fancy, we might imagine the halo of the Milky Way run through with sombre, massive and compact objects, black stars and inky clouds. Just what is needed to excite the curiosity of astronomers. Some have already dreamt up a vast celestial cemetery where stellar corpses and aborted stars gradually accumulate.

However, according to the latest estimates, the fraction of our Galaxy's dark halo that could be explained by baryonic matter (low-luminosity stars and non-luminous compact, massive objects) cannot exceed 20%. These estimates are based on the effect such objects would have on the light from stars in the Magellanic Clouds. It is concluded that the halo of our Galaxy, and probably that of other spirals of this type, is not principally made up of ordinary, atomic matter.

Moving from the galaxies to larger-scale structures, astronomers have used X-ray sensitive telescopes to detect a hot fog in clusters of galaxies. Recent observations by the German satellite ROSAT have shown that the mass of these gases actually exceeds the mass of all the galaxies in the cluster. Moreover, the distribution of the gas can be used to deduce the total mass of the cluster. Dividing by the volume, this leads to an estimate of the density. It is found that cluster densities are around 20% of the critical density. These calculations, based on simple geometrical hypotheses, may nevertheless underestimate the masses of clusters, given the complexity of their structure and internal dynamics.

Summing up the dynamical argument, telescopes reveal the architecture and motions of the cosmos on every scale. Planets, stars, galaxies, and clusters of galaxies are nested one within the other like Chinese boxes. But then the need for dark matter suddenly arises. Without it, stars at the edge of our Galaxy would fly off and the swarms of galaxies in clusters would scatter like birds.

Cosmological densities

Critical density

The critical density is traditionally defined as that density which separates the closed (finite) universe from the open (infinite) universe in the simplest model available, i.e. in a universe without cosmological constant or quintessence. It corresponds to a universe with zero total energy, where the kinetic energy due to expansion is exactly balanced by gravitational potential energy. The value of the critical density is 10^{-29} g cm^{-3}, which amounts to very little when compared to a chunk of iron!

The standard cosmological model predicts three possible futures for our Universe:

1. A continuous and unending expansion if the restoring force due to gravity and hence the density of matter are too small to brake and reverse the general expanding motion. The Universe is said to be open.
2. A global contraction following the current expansion phase, ending in a gigantic collapse if the density of matter is great enough to overcome the general expansion. The Universe is said to be closed.
3. An expansion which slows down forever if the density of matter is just right to exactly balance the general expanding motion. This density is the critical density, of the order of one proton per 10 cubic metres.

Luminous matter density

Systematic analysis of the light from galaxies in a given volume measuring several million light-years in size allows us to specify the present luminosity of the Universe per unit volume. Given the mean colour of the observed sample, a certain mass can be attributed to it by comparison with known stellar populations. For example, a population of solar-type stars emits 2 erg g^{-1} every second, since the luminosity of the Sun is $4 \times 10^{33} \text{ erg s}^{-1}$ and its mass is 2×10^{33} g. From the light collected, we may deduce the light emitted and this in turn allows us to estimate the quantity of radiating matter.

Invisible lenses

Although there can be no doubt about the existence of dark matter, its nature remains to be determined. On the scale of spiral galaxies, the existence of dark matter is inferred by measuring the variation in speed of gas and stars with distance from the galactic centres. On a larger scale, one of the most powerful methods for revealing dark matter is gravitational optics. This seeks out accumulations of matter acting as lenses.

Whole teams of astronomers devote their time to detecting weak lensing or microlensing, that is, the weak focussing effects of massive objects on light. These objects thus serve as gravitational lenses, producing slight distortions in the images of more remote galaxies. It seems that this technique may provide a better window on the invisible matter that astronomers and astrophysicists now consider to be the major component of the Universe (apart perhaps from dark energy). This dark matter is now taken in its immense cosmic dimension.

A French team led by Yannick Mellier and Bernard Fort at the Paris Institute for Astrophysics has used the CFHT (Canada–France–Hawaii Telescope) to identify the colossal superstructure of dark matter along which the galaxies are strung out. The details of this unseeable architecture, its inflections, its reticulations and its fluctuations, conceal invaluable information about the early childhood of the Universe and the growth of its invisible skeleton.

In order to reveal the hidden weave of the cosmos, astronomers have been studying the light emanating from galaxies several billion light-years away. They show up as pale features on the sky. Gravitational effects due to dark matter interposed between them deform their image, slightly flattening their elliptical shape. Neighbouring ellipses point in the same direction, thereby designating the cause of their deformation.

However, these gravitational effects are very slight indeed, and astronomers must eliminate any interference due to imperfections in instrumental optics or atmospheric turbulence.

These measurements are able to breach the otherwise taciturn behaviour of dark matter. Although it refuses to shine, it cannot disguise its gravitational discourse. Astronomers welcome this news with open arms, optimistic that they may succeed in mapping out its contours and determining its weight. In fact, for gravitational astronomy, as for particle physics, to see is to illuminate and analyse the the deflection of the illuminating beam, for it is this deflection which teaches us about the intimate nature of the screen or target. In coming years, gravitational optics will undoubtedly become an effective method for drawing up a large-scale survey of dark matter in the Universe.

Quantitatively speaking, the density of nuclear matter in the Universe is estimated to be between 2% and 5% of the critical density as defined above. The density of visible

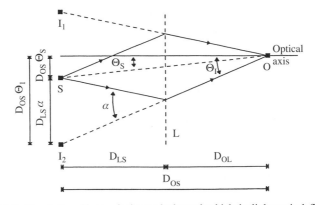

Fig. A1.2. Gravitational lens. α is the angle through which the light ray is deflected. The intrinsic position of the source in the sky is Θ_S, but it is actually observed at Θ_I.

matter, or more precisely, luminous matter, barely reaches 0.5%. What has happened to the difference? It lies in baryonic dark matter. Let us begin by sweeping our own doorstep, so to speak. The problem of nucleonic dark matter is the main motivation for research into gravitational microlensing in the halo of our own Galaxy, as we shall now discuss.

Note that the baryonic density of 2% to 5% is of the same order of magnitude as the galactic density indicated by rotation curves (viz. 5%). It is thus perfectly reasonable to suggest that a large part of galactic matter is located in compact massive objects (CMOs) assumed to swarm around the bright galactic disk. Will these compact massive objects remain forever hidden?

The renowned Polish astronomer Paczynski had the brilliant idea of using gravitational optics to track them down. The plan was to search for gravitational deflection of light from stars in the Magellanic Clouds, nearby galaxies within the gravitational sphere of influence of our own Galaxy.

The invisible deflector plays the role of the optician's glass lens. This is why the term 'lens' was borrowed and combined with the qualifier 'gravitational' to explain the mechanism, that is, to explain exactly what it is that deflects light in this case. The phenomenon involves three protagonists: the observer O, the lensing body L and the source star S. Provided that the lensing body is located at less than one thousandth of a second of arc (0.001 arcsec) from the line of sight of the source star, light rays will be able to reach O by two different paths (Fig. A1.2). Two images are thus superposed, making the source appear brighter than it really is. It is possible to measure this fleeting amplification, which requires close alignment of the three protagonists.

Gravitational (micro) lensing is now recognised as a way of revealing otherwise hidden matter. It is universally used to estimate the distribution and quantity of dark matter on a variety of distance scales. The study of dark matter in the halo of our own Galaxy is not the least significant amongst these.

Compact and dark objects in the galactic halo are expected to produce temporal variations in the apparent brightness of stars located behind them in the Magellanic

Clouds. This occurs due to the focussing of light rays and the relative motions of the observer, lensing body and source. The changing intensity of (unresolved) images caused by the relative motion of the source, lens and observer can be measured. The duration of amplification is a measure of the mass of the lensing body.

If the dark matter in the galactic halo is made of objects with approximately the same mass as the Sun, i.e. Jupiter-sized planets, brown and white dwarfs, neutron stars and black holes, then the passage of such an object across the line of sight to a background star will result in a transient change in its brightness. Relativists will tell you that this fleeting amplification in the apparent brightness of a star when it is almost occulted by a compact massive object is due to a distortion of space–time itself. It is therefore independent of the wavelength of the light, that is, achromatic (not modified by colour changes), and symmetric in its temporal variation. This means that it can be distinguished from the myriad other light fluctuations due to real luminosity changes in certain types of star.

The brightnesses of millions of stars in the Magellanic Clouds have been subjected to close scrutiny. The challenge here is that positive events are extremely improbable. At any given time, only one in every 10 million stars of the Magellanic Clouds is likely to show signs of gravitational amplification.

After eight years spent scouring these two small galaxies, the rewards are slight. Not even a dozen favourable cases have been retained and some of these are contested. This suggests that compact massive objects cannot solve the problem of dark nuclear matter in our Galaxy, or indeed in any galaxy. Can anything be concluded from the few positive sightings? What is the characteristic mass of these objects, given that amplifications last longer for more massive lensing bodies?

To everyone's surprise, no event lasting less than two weeks has ever been observed. The mean mass inferred is half the solar mass, which rules out brown and red dwarfs but favours white dwarfs. However, too many of these remnants from intermediate-mass stars (1–8 $M_\odot$ at birth) would contradict the traditional tenets of astrophysics. Indeed, it would imply a frenzied spate of nucleosynthesis during the formation of the galactic halo. The nature of the compact massive objects thus remains a mystery.

Finally, is it possible to determine the mass fraction of the halo that can be put down to nuclear matter, that is, to matter made up of quarks, since that was our original aim? Arguing from the above observations, this fraction could not exceed 20%, according to the latest estimates by American astrophysicists. This estimate does not disagree with the most cautious reckonings in France.

Consequently, we are forced to admit that compact massive objects are not the main component of the dark halo of our Galaxy, or any of its kind. This in turn implies that the Sun and stars are lost amongst a halo of darkness, in the middle of a haze of neutralinos, hypothetical particles predicted by so-called supersymmetric particle theories.

Nuclear proof

In recent times, several cosmological models have flourished, including one of great renown, known as the Big Bang theory. This is a simple theory with considerable predictive power. The Big Bang is the exceptionally hot and dense state considered to initiate cosmology as we know it today. It explains why stars are invariably made of hydrogen and helium, with only a trace of any heavier elements. Indeed the latter are virtually non-existent in the oldest stars and represent between 2% and 3% of the mass of the youngest stars. The helium content of stars is around 25% by mass and results from a chain of phenomena which took place during the first few minutes of the Universe's long story.

This chain of events involved the so-called weak interaction, a puny and slow force compared with the strong and electromagnetic interactions. The weak interaction governs the conversion of protons into neutrons and vice versa, with creation of a neutrino (antineutrino). It thus determines the lifetime of free neutrons, which naturally decay into protons. In fact, neutrons have a life expectancy of around 10 minutes. However, before they disappear, they may have the opportunity to combine with protons, one which they readily accept. In that case, nuclear physics makes its appearance in the Universe.

At this stage, a remark seems appropriate. If the neutron had precisely the same mass as the proton, it would be stable. We may conclude that the quantity of helium produced during the Big Bang is determined by at least two main facts:

- the inherent instability of the neutron, whose numbers decrease by a factor of two every 10 minutes or so;
- the relative lethargy of the weak interaction, which also explains the long life of the Sun.

The early Universe can be reasonably described as a dilute gas of particles and radiation in thermal equilibrium, uniquely characterised by its instantaneous density and temperature. The expansion of space causes further dilution and cooling of this gas.

The founding phase takes up only a second and yet during this time, the stable components of the material Universe have already emerged, that is, protons, neutrons, electrons and photons. So too have the four distinct forces that will govern them thereafter.

After just one minute, a rush of nuclear reactions takes place, leaving light nuclei amongst its ashes. These include deuterium, helium-3, helium-4 and lithium-7. The amounts of each of the light elements formed during the Big Bang depend crucially on the nuclear density of the Universe, i.e. on the mean number of protons and neutrons per cubic centimetre. This is because light nuclei are created by nuclear reactions, in which dark matter can play no possible role.

By comparing calculated values with the actual content of these various elements in the oldest astronomical objects, we deduce that the density of nuclear matter cannot exceed 5% of the critical density. Now it so happens that the best cosmological theory to date, the theory of cosmological inflation, predicts that the Universe has exactly the critical density. This conclusion is supported by recent observations of remote supernovas and the relic background radiation.

No light is older than the cosmic background, which dates from the time when the Universe first became transparent to light. Tiny variations in temperature have been detected by the COBE satellite and now also by two balloon-borne experiments in the stratosphere, MAXIMA (Millimeter Anisotropy Experiment Imaging Array) and BOOMERANG (Balloon Observations of Millimeter Extragalactic Radiation and Geophysics). Such variations indicate a very slight granularity in the youthful Universe. Cold patches correspond to regions with slightly higher density than the rest. Indeed, such patches shift light towards the red by gravitational effects, making them appear slightly colder. The corresponding regions serve as seeds where future galaxies can condense out. Warmer patches for their part are destined to become immense voids.

The angular diameter of the observed patches is about 1°. Their (real) linear diameter can be estimated as ct, where c is the speed of light and $t = 300\,000$ years is the time when the Universe became transparent. An angular diameter of 1° measured 14 billion years later means that the light rays remained parallel over the whole path and hence that the Universe is globally Euclidean.

It is therefore a Euclidean universe which lies at the meeting point between data from remote supernova studies and observations of the cosmic background radiation. Such a universe contains just enough matter and energy to keep the geometry Euclidean. In fact the Euclidean cosmology fits our Universe like a glove.

Nuclear cosmology: primordial nucleosynthesis

The theoretical edifice of the Big Bang theory rests upon three observations:

- the uniform distribution of matter on large scales and the isotropic expansion that preserves it;
- the existence of the almost uniform and exactly thermal background known as the cosmic microwave background;
- the abundances relative to hydrogen of the light elements deuterium, helium-3, helium-4 and lithium-7.

The recession of the galaxies attests to the general idea that the Universe is expanding. The cosmological background provides indisputable evidence of a hot, dense beginning. Finally, the existence of the lighter atoms in the measured proportions is witness to very early nuclear synthesis.

Primordial nucleosynthesis really puts the Big Bang cosmology to the test. One might call it a baptism of fire. From these brief but brilliant and fertile beginnings arose a series of light nuclei that are today found everywhere in nature: above all hydrogen, followed by helium, which between them amount to 98% of the total mass of atomic matter in the Universe.

The basic parameters of this problem are the lifetime of the neutron (887 seconds) and the number of neutrino species (three), both given by modern microphysics. At the time which interests us here, i.e. $t = 1$ s, the energy density of electromagnetic radiation was greater than that of matter. This is therefore referred to as the radiation era.

As the weak interaction is the slowest of all, it was the first to find itself unable to keep up with the rapid expansion of the Universe. The neutrinos it produces, which serve as an indicator of the weak interaction, were the first to experience decoupling, the particle equivalent of social exclusion. By the first second, expansion-cooled neutrinos ceased to interact with other matter in the form of protons and neutrons. This left the latter free to organise themselves into nuclei. Indeed, fertile reactions soon got under way between protons and neutrons. However, the instability of species with atomic masses between 5 and 8 quickly put paid to this first attempt at nuclear architecture. The two species of nucleon, protons and neutrons, were distributed over a narrow range of nuclei from hydrogen to lithium-7, but in a quite unequal way.

The crucial species in primordial nucleosynthesis of the light elements is helium. There are two reasons:

- it has an extremely stable nucleus and many nuclear reactions lead to it, short-circuiting neighbouring nuclei;
- it has exceptionally unstable progeny (nuclei with masses between 5 and 8), so that the production line breaks down at mass 7.

It is only much later that the stars pick up the task of nuclear complexification, favouring triple mergers of the type $^4He \rightarrow {}^{12}C$, something which the Big Bang was quite incapable of putting into practice due to the rapid dilution effect occurring at the time.

A proton and a neutron together make a nucleus of deuterium, an unavoidable step in the chain of nuclear reactions. At the beginning, this fragile nucleus would have been bathed in an ocean of high-energy photons. These photons would immediately have dissociated it, before it could find another proton with which to join up and form less fragile species made up of three nucleons, like tritium ^3H and helium-3. However, the temperature was falling. At some point, when the temperature had fallen to 1 billion K, deuterium nuclei would suddenly be able to resist the much less aggressive photons circulating at these temperatures. Constructive relations could then be established with other particles in the primordial soup. Having thus escaped from the tyranny of the photons, it would have given itself over to the nuclear construction project, burning almost completely away to produce helium-3, helium-4 and a dash of lithium-7.

For a given neutron, the probability of taking part in a nuclear combination is proportional to the proton density. High densities favour marriage, whilst low densities lead inescapably to death.

It is easy to see why the results of primordial nucleosynthesis, and in particular the final abundance of deuterium, should be so sensitive to the nucleonic density of the Universe. For this reason, deuterium, made up of one proton and one neutron, can be considered as an excellent cosmic densimeter. The disparate abundances for their part are related to specific nuclear properties of the isotopes under consideration.

Quantitative predictions of the theory concerning naturally occurring proportions of nuclei with masses 1, 2, 3, 4 and 7 can then be compared with the actual proportions measured in the oldest and most distant astronomical objects (stars and gas clouds) (Fig. A1.3). Agreement between prediction and observation is found by fitting the only free parameter of the theory, that is, the nucleonic density of the Universe today (which is easily related to its counterpart at the time of the Big Bang). In fact, agreement is obtained if and only if the value obtained here is between 1 and 3×10^{-31} g cm^{-3}. This is indeed an astonishingly accurate value when we consider that it refers to an event occurring some 15 billion years ago! It thus transpires that the baryons contribute at most 5% of the critical density.

A single parameter, the nucleonic density, is thus sufficient to explain the proportions of the light elements in the Universe, from helium at 10% of the number of hydrogen atoms to lithium at one ten-thousandth. However, the number of neutrino species must be at most three in order to avoid an overproduction of helium-4. Since each neutrino belongs to a single particle family, the number of particle families in the Universe must

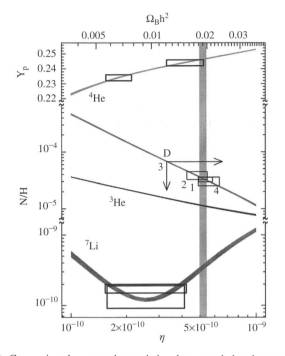

Fig. A1.3. Comparison between observed abundances and abundances predicted by the theory of primordial nucleosynthesis. The horizontal axis shows the ratio η between the number of baryons and the number of photons. The vertical axis shows the mass fraction of helium and the numerical ratios D/H, ^{3}He/H and ^{7}Li/H. Observational data are represented by boxes with height equal to the error bar. In the case of helium and lithium, there are two boxes, indicating the divergence between different observers. Deuterium holds the key to the mystery, but it is difficult to measure. The region of agreement is shown as a shaded vertical ribbon (after Burles & Tytler 1997). A higher level of deuterium would lead to a lower baryonic density, of the order of 2%. This would agree better with the lithium data, which have been remarkably finely established. This idea is supported by E. Vangioni-Flam and shared by myself. (From Tytler 1997.)

also be exactly three. (The first family contains the up and down quarks, denoted u and d, the electron and the electron neutrino. The second contains the strange and charmed quarks, denoted s and c, the muon and the muon neutrino. The third contains the bottom and top quarks, denoted b and t, the tauon and the tauon neutrino. The material world as we know it is made up exclusively of the first particle family.)

Dark matter

A whole series of conclusions follow from comparison between the density of nuclear matter and other densities determined by theory (T) or observation (O). These can be summarised in the following way:

- the critical density (T), i.e. that density which is just sufficient to slow down the expansion of the Universe and bring it to a halt after an infinite time, at least if the cosmological constant is zero;
- the density of luminous matter (O);
- the density of gravitating matter (O + T) as determined by studying the dynamics of clusters of galaxies;
- the cosmic density (O + T), i.e. the total matter and energy density of the Universe in all existing forms, equal to the critical density if we are to believe the cosmological inflation model which currently holds sway, and data relating to distant supernovas and the cosmic microwave background.

An inventory of the Universe

Critical density $= 10^{-29}$ g cm^{-3}

density of luminous matter/critical density $= 0.005\%$

density of gravitating matter/critical density $= 10–30\%$

density of nuclear matter/critical density $= 2–5\%$

total density of the Universe/critical density $= 1$.

Note that the density of nuclear matter is greater than the density of luminous matter, but less than the density of gravitationally active matter. Note also that it is very much smaller than the critical density. We may therefore deduce the existence of dark nuclear matter and dark (or even invisible) non-nuclear matter.

Let us examine this situation in more detail. It is quite clear that the density of matter in clusters of galaxies is significantly higher than the density of nuclear matter as deduced from primordial nucleosynthesis (2–5% of the critical density). If we assume that these structures are representative of the Universe as a whole, then in order to make up the difference, we are forced to resort to clouds of exotic elementary particles left over from the Big Bang. The fate of the Universe then lies in the hands of non-nuclear matter of unknown but not unknowable nature (e.g. neutralinos).

Furthermore, if we assume that the total density of matter and energy equals the critical density, for theoretical and aesthetic reasons related to the inflationary theory of the Universe (Cassé 1993), we are forced to the conclusion that there is another form of matter. In other words, apart from the dark matter already mentioned, which contributes at most 30% of the critical density, there exists another form of matter just as transparent as the first, making up about 70% of the Universe. This conclusion seems to be corroborated by observations of distant supernovas, which suggest that the expansion of the Universe is actually accelerating, whereas it was always expected to be decelerating. This acceleration is imputed to the properties of a certain material (energetic) substrate which produces a repulsive gravitational effect. However, the observations in question are still too recent to be taken as a firm proof. Let us call this new form of matter the quintessence, in homage to the ancient philosophers. Whilst we await further corroboration, let us simply note that all that shines is matter, but that not all matter shines, and that there exists a certain species of antigravitational matter.

Finally, we may divide up the cosmos in the following way:

$$\text{cosmos} = \text{quintessence} + \text{matter},$$
$$\text{matter} = \text{non-nuclear matter} + \text{nuclear matter},$$
$$\text{nuclear matter} = \text{dark nuclear matter} + \text{luminous nuclear matter}.$$

The dominant form has been placed first on the right hand side of each line. The ratios between the two terms of each line are roughly as follows:

$$Q/M = 2.3,$$
$$NNM/NM = 10 \text{ to } 30,$$
$$DNM/LNM = 5.$$

Abstractors of quintessence

The theory of inflation is an attempt to extend the Big Bang theory back as close as possible to zero time. The main hypothesis is a period of ultra-rapid expansion during the first fraction of a second. This expansion of space is induced by a negative pressure. With a change in the sign of the pressure, the world is turned upside down!

The inflationary process, resulting in an exponential expansion of the Universe right at the beginning of its history, is designed to explain the Euclidean nature of the Universe and the remarkable uniformity of the cosmological background radiation (Cassé 1993; Reeves 1996). The theory posits a substrate with negative pressure at the beginning of the Universe. It accomplishes its explosive mission and disappears in favour of light and matter.

According to theories which assume the existence of some form of matter with repulsive gravity (antigravity), the Universe possesses an accelerator (an antigravitational substance) in addition to matter with attractive gravity (luminous and dark matter) which produces a braking effect. Quintessence would be the residue from this primordial repulsive substance and it would retain its attributes.

Even recently, it was thought that the history of the Universe was simply an expansion working against the opposing forces of gravity. Indeed, gravity has a braking effect on such expansion, playing the role of a restoring force. In stark contrast, the effect of quintessence is to ascribe a repulsive attribute to space on a very large scale, thereby accelerating the expansion. As a result, the expansion rate measured today would be greater than that in the past. The age of the Universe would then be increased.

Is there any tangible evidence for these claims? Some would answer in the affirmative. Among them is Saul Perlmutter at the Lawrence Berkeley Laboratory in Berkeley, who leads the Supernova Cosmology Project, and Brian Smith, a member of the Mount Stromlo and Siding Springs observatories (Australia) and mentor of the High-z Supernova Search Team. Their point of common interest is the most distant type Ia supernovas ever observed.

As we have already seen, the type Ia supernovas occur in tight binary systems containing a white dwarf. The latter tears matter away from its bulkier companion, fattening itself up until it exceeds a certain critical mass. The whole star is then destroyed in a gigantic nuclear deflagration. The main feature of type Ia events is high luminosity and extreme regularity in their characteristics. For this reason, they serve as standard candles, marking out the vast reaches of the cosmos.

The explosion mechanism remains somewhat mysterious, as already explained, but the nature of their progenitor seems to have been clearly established. The culprits are greedy white dwarfs which, having eaten up a portion of their companion star, go over the critical mass of 1.4 solar masses beyond which no white dwarf can survive. Fusion of carbon at high densities then takes on an explosive turn. The nuclear energy released, some 2×10^{51} erg, easily exceeds the 10^{50} erg gravitational binding energy of the white dwarf, which promptly evaporates in the explosion. The neutrino flux emitted in such events is quite insignificant.

The deceased departs in light, with no flower, nor crown, nor neutrinos, nor neutron star. And what a light it is! A single type Ia supernova shines for a few days as brightly as a billion suns, in other words, as brightly as a small galaxy. Nuclear incineration of carbon and oxygen produces a considerable mass of nickel-56 which transmutes into iron-56, releasing gamma rays at a precise energy, i.e. nuclear gamma lines. This radioactivity is the energy supply of the supernova.

These small, sturdy, carbon- and oxygen-bearing white dwarfs deliver a clear message by their explosion. They can be picked out from the crowd by their extreme luminosity, and their highly specific spectra and light curves. The huge potential of type Ia supernova events has been known for some time, but it was only during the 1990s that supernova research itself exploded upon the world.

Following one of the biggest inquiries ever held in modern astronomy, it transpires that their apparent luminosity is slightly less than would be found if space were Euclidean and expansion were merely slowed down by the gravitational effects of matter. In fact, the expansion is more vivacious than was previously thought. This means that distances to remote objects are slightly distended, so that the supernovas appear less luminous than expected.

This investigation, involving the most up-to-date instruments such as the Hubble Space Telescope, led to the following conclusion: the data disagree strongly with the hypothesis that the Universe is flat and contains no quintessence, thus ruling out one of the most favoured cosmological models.

Accepting this result amounts to saying that the expansion of the Universe has been speeding up for several billion years. This may seem surprising, since all matter, including dark matter, always works to slow down such a movement. It would thus seem that the Universe has begun another inflationary phase! If this were the case, the history of the Universe could be divided into four reigns, with transfer of power between the first quintessence, radiation, matter, and finally, the second quintessence. Expansion would be alternatively gentle and violent and the era of the second quintessence irreversible.

This conclusion destroys any hope of eternal return on the cosmic scene. However, it only applies if we can assume that past supernovas were exactly like today's, that is, those observed in nearby galactic suburbs. This assumption has raised some suspicion amongst specialists in stellar evolution.

On the other hand, we must somehow close the Universe, or more precisely, find some way of giving it the critical density, since this is what inflation demands. Indeed, it is required not only by inflationary theory, but also by close scrutiny of the leopard skin pattern that constitutes the microwave background, radiative relic from the Big Bang. We may thus conclude that, even if the American and Australian results are contested, they do bring out a basic truth, although maybe purely by chance. More and more scientists now assume that the quintessence accelerates expansion of the Universe and accounts for 70% of the world's balance sheet.

The question will certainly be decided in 2007 when the PLANCK satellite is launched. This showpiece of French and European astrophysics is destined to carry out an extremely refined inventory of the Universe, determining the contribution from nuclear matter, dark matter and quintessence. It will do this by studying once again the cloak of light cast over the Big Bang, the cosmological background radiation, but this time peering within its very finest folds (Lachièze-Rey & Gunzik 1995; Silk 1997).

The contents of the Universe as estimated by dynamical or nuclear criteria are not sufficient to reach the critical density, as required by inflation. And no more need be said about the worthiness of the inflationary theory. The quantity of luminous matter assembled within the galaxies is less than the total quantity of nuclear matter as deduced from cosmological arguments. Where is the difference between luminous and universal matter? In the invisible, of course. But this is the worst of all invisibilities. For it is not due to the limitations of the retina, but rather to the mute inarticulacy of certain substrates which resolutely refuse to shine. The great majority of matter and energy is dark, atonal with respect to the registers of light.

The insufficient density of ordinary matter (i.e. yours, mine and the matter in stars) implies that there must be matter which is not nuclear, literally extraordinary matter which is not made up of atoms. The sky thus distances itself from the purely atomic and nuclear paradigm. So speaks the dark side of space.

Appendix 2

Supernovas and cosmology

Glossary

Hubble redshift–magnitude diagram an important cosmological tool expressing the rate of expansion of the Universe in the past
light curve changing luminosity of a celestial object as a function of time
magnitude measure of the brightness of celestial bodies. A difference in luminosity by a factor of 100 corresponds to a difference of five magnitudes. The magnitude decreases as the brightness of the object increases. The apparent magnitude of an object characterises its brightness as seen from Earth. Its absolute (intrinsic) magnitude is the magnitude it would have if it were placed at a standard distance of 10 parsecs from Earth. The absolute magnitude allows for an objective comparison of the luminosity of different stars and galaxies.

Properties of supernovas

At first glance, the spectral properties, absolute magnitudes (intrinsic luminosities) and shapes of the light curves of the majority of type Ia supernovas (SNIa) are remarkably similar. Only a few rather subtle photometric and spectrometric differences can be discerned from one object to another.

Hydrogen shines by its absence and the optical spectra of SNIa events feature spectral lines of neutral and once ionised elements (Ca^+, Mg^+, S^+ and O^+) at the minimum of the light curve. This indicates that the outer layers are composed of intermediate mass elements. SNIa events reach their maximum luminosity after about 20 days. This luminous peak is followed by a sharp drop amounting to three magnitudes per month. Later the light curve falls exponentially at the rate of one magnitude per month.

The exploding stars discussed here of a much more modest character than gravitational collapse supernovas, which arise when the core of a gigantic star gives way. However, they eject matter at very high speeds, up to $10\,000$ km s^{-1}, and release a comparable amount of energy, some 10^{51} erg. They are genuine nuclear bombs, unlike the gravitational-collapse supernovas (SNII), which draw their energy from gravity. (Supernovas are classified under the headings SNIa, SNIb, SNIc and SNII. Those carrying the letters b and c should actually be grouped together with SNII in the gravitational-collapse category. The only point they have in common with SNIa is the absence of

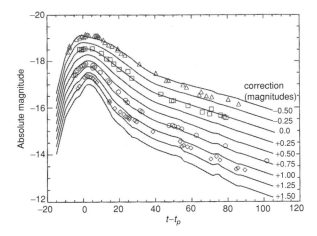

Fig. A2.1. Light curves for various SNIa events. The figure shows empirical families of light curves. The brightest shine longer, at the peak of their glory. (From
Riess *et al.* 1998.)

hydrogen features in the spectrum. However, this absence is explained by the fact that
their envelope has evaporated or been torn off by a close companion.)

The most striking feature of the SNIa events is thus that they all look remarkably
alike, a characteristic not shared by other supernova categories. Their spectral evolution
repeats itself from one such event to the next. Apart from a few black sheep, SNIa
light curves show an impressive uniformity. The scatter in absolute magnitude values is
about 0.2 magnitudes over the whole sample so far observed, well below the magnitude
variation shown by supernovas of other types. This is why SNIa events are considered
to be a good standard candle which may be used to calibrate absolute luminosities on a
cosmological scale.

The important datum for cosmology is precisely the luminosity at the peak of the light
curve. It is crucial to be able to establish this maximum value in order to use the SNIa
event as a distance indicator. Correctly calibrated and reproducible light curves from
type Ia supernovas have become a major tool for determining the local expansion rate
and geometrical structure of the Universe (Fig. A2.1). A great deal of effort has been
put into producing adequate models of these events over the past few years.

Through a well-established tradition, relatively close SNIa events ($z < 0.1$) have
been used to measure the current local value of the Hubble parameter. Supernova-based
cosmology has seen a recent upturn in activity. The local expansion rate is something
like 60 km s^{-1} per megaparsec, which corresponds to 18 km s^{-1} per million light-years.

Systematic research using wide-field images taken at intervals of three to four weeks
have allowed two independent groups, the Supernova Cosmology Project and the High-z
Supernova Search Team, to identify more than 50 SNIa events at intermediate redshifts.
The Hubble redshift–magnitude diagram has been extended out to $z = 1$.

Today, a certain level of diversity has been discovered in SNIa events. Before they can
be used as cosmological distance standards, it must be checked that their maximal luminosity has an appropriate value. Otherwise *ad hoc* corrections must be brought to bear to

homogenise the values. To this end, any systematic difference from the norm is sought out and deviants excluded. Here again, theory is called in to distinguish pathological behaviour from a perfectly normal scatter about the average.

Many connections have been found between the luminosity peak, the shape of the light curve, evolution in the colour, spectral appearance, and membership of a galaxy of given morphology. However, after the first 150 days, uniformity takes over and all these objects fade in the same way and with the same spectrum.

Roughly speaking, SNIa events can differ in their explosive power. It is observed that supernovas with the least violent explosion are less bright and redder in colour. Furthermore, they fade more quickly and eject matter more slowly than the violent explosions. The relationship between the width of the light curve (its duration) about the maximum and the peak brightness is the most significant of these correlations. It has been used to calibrate the peak brightness of a range of different SNIa events and to substantially reduce the scatter in absolute brightness.

This correction plays a key role in any cosmological application. Without it, SNIa events could not be used as distance indicators. However, its purely empirical nature remains unsatisfactory to demanding theoretical minds. We would like to be able to explain physically why some explosions are weaker than others, and what effect this has on the appearance of the object. This involves building detailed models of these explosions and the way radiation is transferred through the expanding envelope, similar to those made to describe atomic bombs or spheres struck by laser beams, which implode before exploding.

Observed empirical correlations, such as the one relating the width of the light curve to its peak brightness, are thus used to bring the images of these supernovas into line, making them all look alike. An archetypal or standard type Ia supernova is thereby defined. But this systematic correction, however shrewd and useful it may be, will never totally satisfy the astrophysicist, always seeking out a finer understanding of these phenomena.

For theoreticians, such developments provide an opportunity to refine and test their more subtle inventions and to explain apparent correlations. I am quite sure they will seize this particular opportunity, given the extravagant amount of data already accumulated, to choose amongst a crowd of competing models the one which best accounts for reality.

Let us now turn our attention to the relationship between stellar physics (extended to supernovas) and observational cosmology. Edwin Hubble is remembered for the discovery that the Universe is expanding. Less well known is the redshift–magnitude plot that carries his name. Magnitude is a measure of the apparent brightness of celestial objects, and redshift is a measure of their recession speed and hence their distance.

The redshift–magnitude or $z–m$ diagram (Fig. A2.2) has become a standard tool in cosmology because it can be used to determine past variations in the expansion rate, through what is usually called the deceleration parameter. Any acceleration or slowdown of the expansion can thus be brought to light.

The analysis is based upon 42 SNIa events with redshifts between 0.18 and 0.83, combined with 18 nearby SNIa events with $z < 0.1$. These were discovered in the context of a systematic investigation, the Calan/Tololo Supernova Study. The distances of those supernovas with the highest redshift are on average 10–15% greater than would be obtained in the case of a low-density universe ($\Omega_m = 0.2$) with no quintessence.

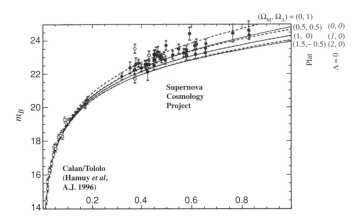

Fig. A2.2. Redshift–magnitude diagram: z against m. (From Perlmutter *et al.* 1999.)

The cosmological parameters Ω_m and Ω_Λ can be determined from the Hubble diagram, provided that we have some well-calibrated cosmological standard candle that can be observed across a wide range of redshifts. This is precisely the approach adopted by Riess *et al.* (1998) and Perlmutter *et al.* (1999) when they used the type Ia supernovas. Pilar Ruiz-Lapuente at the University of Barcelona and Renald Pain at the University of Paris both made contributions to this exemplary cosmological programme.

A distance measure d_L known as the luminosity distance is relevant here. In a spatially flat space–time, it is given by the formula

$$\frac{H_0 d_L}{c(1+z)} = \int_0^z \left[(1+z')^2(1+\Omega_m z') - z'(2+z')\Omega_\Lambda\right]^{-1/2} dz'.$$

The distance, and hence the apparent magnitude, is sensitive to both Ω_m and Ω_Λ. At small z, the expression reduces to the well-known Hubble law. However, at large z, the deviation from Hubble's law grows ever larger. This is why distant SNIa events are so significant in choosing between cosmological models.

The two American groups arrive at the values $\Omega_m \approx 0.3$ and $\Omega_\Lambda \approx 0.7$. Once our initial excitement has died down, it is important to note that this result is far from conclusive. The reason is that the measured effect is extremely small, being of the order of 0.25 magnitudes, and can be attributed to other causes:

- the Hubble parameter H_0 may have a high level locally;
- supernovas evolve, like any other object in the Universe, and despite systematic corrections, this effect may have been only partially eliminated;
- light from supernovas may be focussed or defocussed by gravitational lenses.

These are only the most obvious sources of distortion.

For most astronomers, the solution to these cosmological problems resides in a combination of various methods. The luminosity–redshift test must be combined with independent techniques, such as anisotropies in the cosmic background radiation and statistical study of gravitational lenses.

Only when all the data converge upon a common answer will we be able to say that the problem of the cosmological parameters has been resolved. And although we have now started off along this road, some patience is still required. Here again, theory will be brought to bear, not only on the normal or average individuals, but also on the deviants. For whatever distinguishes appearances has a deep foundation. We cannot be satisfied with empirical or phenomenological relationships when reducing light curves to a common model. This means that the theoretical modelling of binary systems ending their existence in a cataclysmic explosion remains one of the central aims of astrophysics.

Appendix 3

Explosions

Glossary

neutron excess the difference $N - Z$ between the number of neutrons and the number of protons in a nucleus

Nuclear statistical equilibrium

The most tightly bound nuclei, i.e. the most stable and robust, in the iron peak are not symmetric arrangements bringing together equal numbers of protons and neutrons ($N = Z$). Rather, they possess a neutron excess ($N - Z$) between 2 and 4. Close to iron, the most stable nucleus ^{56}Fe has a number of neutrons which exceeds the number of protons by 4 units ($N - Z = 4$).

The isotopic and elemental abundance table shows that, in the Solar System, iron is more abundant than its neighbours. Analysis of stellar spectra confirms this result, giving it a universal character.

Theoretically, nuclear strength is enhanced by internal transmutations of protons into neutrons, under the mandate of the weak interaction, either by positron emission (p → n + e$^+$ + ν) or by electron capture (p + e$^-$ → n + ν). However, the weak interaction is much slower than the strong interaction. The question remains as to whether it will happen inside the star, or outside, once the matter has been expelled, i.e. after the explosion. This is not just an academic question. The answer we give will determine whether or not we can corroborate explosive nucleosynthesis by observation.

Any attempt to understand the conditions in which iron and its kin were created, and identify the astrophysical site of their birth, must focus on the idea of nuclear statistical equilibrium. The situation is the exact nuclear analogy of the ionisation equilibrium occurring in hot gases.

The abundance of each element is fixed by its binding energy, which characterises its strength as an entity, and the temperature and density of free neutrons and protons attacking the nucleus (Fig. A3.1). If, as is usually the case, nuclear equilibrium is reached before a significant number of radioactive decays have had the time to occur, an auxiliary constraint can be imposed: the total number density of protons and neutrons, both free and bound, must preserve the mean n/p ratio.

A small growth in the n/p ratio has a considerable effect on the composition of the iron peak. For values very close to unity, the most abundantly produced isotope in the absence

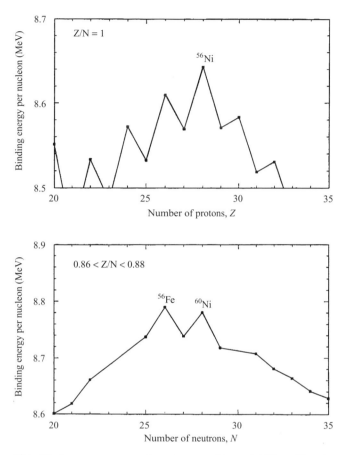

Fig. A3.1. Binding energy per nucleon in symmetric nuclei ($Z = N$) and asymmetric nuclei ($0.86 < Z/N < 0.88$). ^{56}Ni is the most tightly bound nucleus with an equal number of protons and neutrons, whilst ^{56}Fe is the strongest nucleus with $Z/N = 0.87$. Nuclear statistical equilibrium favours ^{56}Fe if the ratio of neutrons to protons is 0.87 in the mixture undergoing nucleosynthesis. In fact nature seems to have chosen to build iron group nuclei in a crucible with $Z = N$.

of radioactive β decay is nickel-56. Amongst the symmetric isotopes ($Z = N$), this is the one with the greatest binding energy per nucleon. The predominance of iron-56 in nature leads to the conclusion that the synthesising process happens so quickly that the main nucleus produced is nickel-56, which then decays to iron. This process has many consequences. Indeed, it explains supernova light curves.

We know today that nuclear statistical equilibrium in a neutron-poor environment (p/n = 1.01), dominated by nickel-56 rather than iron-56, gives a good overall explanation of the abundance table in the neighbourhood of the iron peak. This is a natural consequence of high-temperature combustion. The corresponding combustion times are so short that the sluggish weak interaction just does not have the time to influence matters. In fact the n/p ratio in the core of the star, resulting from the whole of its nuclear history, is not modified. Explosive combustion clips the wings of the weak interaction.

Science of heat

The critical factor which thus determines the nature of the emerging iron peak is the combustion time. Only a hydrodynamic model can provide us with a full appreciation of fusion times, describing the sudden and brutal temperature increase as the shock wave passes through, and the equally sudden cooling that follows it.

The first attempt to calculate nucleosynthesis during the passage of a shock wave through the silicon-rich region, carried out by Truran, Arnett and Cameron in 1967, has been largely confirmed. The most abundantly synthesised nuclei in these conditions turn out to be nickel-56 and the unstable α nuclei beyond calcium (titanium-44, chromium-48, iron-52 and zinc-60). This is perhaps the most beautiful result in the whole theory of nucleosynthesis, for it shows that iron, used in railways and haemoglobin, the king of nuclear creation, one might say, is not actually produced as iron, but in the form of radioactive nickel.

It also shows just how sensitive are the results of explosive nucleosynthesis to the n/p ratio. A numerical experiment should convince us of this:

1. Take a pure sample of silicon-28 ($N = Z$) at a density of $2\,000\,000$ g cm^{-3}, heat it to 3 billion k and let nuclear evolution have its way until all the silicon has disappeared. We observe a gradual growth of iron-56, leading to its final domination.
2. Now do the same but switching off the weak interaction. This time we obtain a very different behaviour, because the dominant species is nickel-56. The gas retains its egalitarian nature, i.e. $Z = N$.
3. Finally, consider a temperature of 4 billion k rather than 3 billion, and a density of $20\,000\,000$ g cm^{-3}, but leave the weak interaction switched on. We find that nickel-56 once again predominates over other iron peak nuclei, and this very rapidly indeed.

The composition of the ashes from the nuclear fuel (in this case, silicon, characterised by n/p $= 1$) is a direct consequence of the degree of neutron enrichment allowed by β decay in the time available for nucleosynthesis, which is itself fixed by astrophysical conditions.

A full appreciation of the combustion time requires:

- a model providing the configuration of the star just before it explodes;
- a hydrodynamic model able to follow the temperature, density and composition of the matter during the explosion;
- knowledge of nuclear and weak interaction rates for all nuclear species taking part in the silicon transmutation process under the effects of intense heat and in the context of nuclear statistical equilibrium.

In a virtual sense, we observe the behaviour of nuclei in the nuclear mixture as others would observe the behaviour of cells in an incubator.

It is a straightforward matter to understand the physical constraint imposed by the combustion time of silicon. Ashes from silicon will be all the richer in neutrons as the weak interaction is given the time to carry out its work via positron emission and electron capture, changing protons into neutrons and hence enriching the environment with neutrons.

At very high temperatures, above 3 or 4 billion k, silicon is consumed so quickly that positron emission and electron capture reactions which might modify the n/p ratio are largely short-circuited. The weak interaction does not have time to convert any appreciable fraction of protons into neutrons during the brief period of thermonuclear combustion. It follows that, starting with matter that is initially dominated by nuclei containing equal numbers of neutrons and protons, such as oxygen-16 and silicon-28, the final products must conserve $Z = N$, unless they move away from nuclear stability beyond calcium-40, the last stable α element.

The dominant nuclear species resulting from processes at temperatures between 4 and 6 billion k include titanium-44, chromium-48, iron-52 and 53, nickel-56 and 57 and zinc-58, 60, 61 and 62. Isotopic abundances resulting from radioactive β^+ decay of these nuclei are compatible with terrestrial and meteoritic measurements relating to calcium-44, titanium-48, chromium-52 and 53, iron-56 and 57, and nickel-58, 60, 61 and 62.

The proportions with respect to iron of vanadium-51, manganese-55 and cobalt-59, formed as manganese-51, cobalt-55 and copper-59, respectively, also agree with abundances observed in the Solar System. The overall coherence of this behaviour suggests that appropriate conditions for synthesis of iron peak elements should be sought in supernovas.

Furthermore, the fact that iron is synthesised in the form of a nickel isotope has important implications from an observational standpoint. Indeed, it provides a check on the foundations of the whole theory of explosive nucleosynthesis. These implications are twofold, as we have seen. They concern supernova light curves and gamma emission from these objects.

Appendix 4
Stellar nucleosynthesis

Glossary

yield quantity of various isotopes synthesised and ejected by a star of given mass and metallicity; the nuclear donation of each star

Nuclear evolution

In the twentieth and twenty-first centuries, humankind has turned its attention not only to the energy of matter and attempts to master it, but also to the origin and evolution of the elements that make it up.

Nuclear evolution precedes and determines the evolution of life, and is itself preceded by the evolution of elementary particles. Such is the great scheme of material things. The idea of universal unity can only be strengthened by this knowledge. In this physical genesis, the star plays a crucial intermediate role between the Big Bang and life, and for this reason we owe it our closest attention.

The birth certificate of any member of the stellar society carries one of the three following comments:

- alone or accompanied;
- mass;
- metallicity.

Masses range between 0.1 and 100 $M_\odot$ and metallicities from one-thousandth of the solar value up to the latter. Metallicity is a question of generation. It may be taken as our leitmotif that the most ancient of stars are also the poorest in metals.

This therefore defines the stellar condition. Binary stars behave differently to single stars, as is clear from the case of the type Ia supernovas. In order to simplify, let us leave them aside for the moment and consider the case of the only child. For isolated stars, knowledge of the two other attributes, mass and metallicity, suffice to characterise its nuclear opus, that is, the amounts of the various elements that it will eventually release into space, as well as its lifespan.

Stars do not all work at the same rate, and nor do they produce the same nuclear species. Depending on its mass, and to a lesser extent its metallicity at birth, each star delivers its specific batch of atoms to the surrounding region of space, thus making its

Table A4.1. *Yields of different stars with masses between 13 and 70 $M_\odot$*

Isotope	13 $M_\odot$	15 $M_\odot$	18 $M_\odot$	20 $M_\odot$	25 $M_\odot$	40 $M_\odot$	70 $M_\odot$
^{16}O	1.51×10^{-1}	3.55×10^{-1}	7.92×10^{-1}	1.48	2.99	9.11	2.14×10^{1}
^{18}O	9.44×10^{-9}	1.35×10^{-2}	8.67×10^{-3}	8.68×10^{-3}	6.69×10^{-3}	1.79×10^{-6}	3.80×10^{-3}
^{20}Ne	2.25×10^{-2}	2.08×10^{-2}	1.61×10^{-1}	2.29×10^{-1}	5.94×10^{-1}	6.58×10^{-1}	2.00
^{21}Ne	2.08×10^{-4}	3.93×10^{-5}	2.19×10^{-3}	3.03×10^{-4}	3.22×10^{-3}	2.36×10^{-3}	1.14×10^{-2}
^{22}Ne	1.01×10^{-4}	1.25×10^{-2}	2.74×10^{-2}	2.93×10^{-2}	3.39×10^{-2}	5.66×10^{-2}	5.23×10^{-2}
^{23}Na	7.27×10^{-4}	1.53×10^{-4}	7.25×10^{-3}	1.15×10^{-3}	1.81×10^{-2}	2.37×10^{-2}	6.98×10^{-2}
^{24}Mg	9.23×10^{-3}	3.16×10^{-2}	3.62×10^{-2}	1.47×10^{-1}	1.59×10^{-1}	3.54×10^{-1}	7.87×10^{-1}
^{25}Mg	1.38×10^{-3}	2.55×10^{-3}	7.54×10^{-3}	1.85×10^{-2}	3.92×10^{-2}	4.81×10^{-2}	1.01×10^{-1}
^{26}Mg	8.96×10^{-4}	2.03×10^{-3}	5.94×10^{-3}	1.74×10^{-2}	3.17×10^{-2}	1.07×10^{-1}	2.91×10^{-1}
^{27}Al	1.04×10^{-3}	4.01×10^{-3}	5.44×10^{-3}	1.55×10^{-2}	1.95×10^{-2}	8.05×10^{-2}	1.44×10^{-1}
^{28}Si	6.68×10^{-2}	7.16×10^{-2}	8.69×10^{-2}	8.50×10^{-2}	1.03×10^{-1}	4.29×10^{-1}	7.55×10^{-1}
^{29}Si	7.99×10^{-4}	3.25×10^{-3}	1.76×10^{-3}	9.80×10^{-3}	6.97×10^{-3}	5.43×10^{-2}	1.08×10^{-1}
^{30}Si	1.87×10^{-3}	4.04×10^{-3}	3.33×10^{-3}	7.19×10^{-3}	6.81×10^{-3}	4.32×10^{-2}	1.00×10^{-1}
^{31}P	2.95×10^{-4}	6.55×10^{-4}	4.11×10^{-4}	1.05×10^{-3}	9.02×10^{-4}	5.99×10^{-3}	2.57×10^{-2}
^{32}S	1.46×10^{-2}	3.01×10^{-2}	3.76×10^{-2}	2.29×10^{-2}	3.84×10^{-2}	1.77×10^{-1}	2.05×10^{-1}
^{33}S	1.19×10^{-4}	9.60×10^{-5}	1.48×10^{-4}	8.84×10^{-5}	2.20×10^{-4}	7.49×10^{-4}	1.02×10^{-3}
^{34}S	1.83×10^{-3}	1.49×10^{-3}	1.89×10^{-3}	1.26×10^{-3}	2.77×10^{-3}	1.14×10^{-2}	1.98×10^{-2}
^{35}Cl	3.70×10^{-5}	3.45×10^{-5}	8.95×10^{-5}	6.05×10^{-5}	6.72×10^{-5}	4.75×10^{-4}	1.76×10^{-3}
^{37}Cl	6.73×10^{-6}	9.60×10^{-6}	1.04×10^{-5}	4.96×10^{-6}	1.32×10^{-5}	1.17×10^{-4}	1.01×10^{-4}
^{36}Ar	2.36×10^{-3}	5.63×10^{-3}	6.13×10^{-3}	3.78×10^{-3}	6.71×10^{-3}	3.11×10^{-2}	2.92×10^{-2}
^{38}Ar	4.85×10^{-4}	6.49×10^{-4}	6.29×10^{-4}	3.25×10^{-4}	7.24×10^{-4}	9.14×10^{-3}	6.16×10^{-3}
^{39}K	1.95×10^{-5}	3.31×10^{-5}	3.66×10^{-5}	3.24×10^{-5}	3.47×10^{-5}	3.83×10^{-4}	3.84×10^{-4}
^{41}K	1.42×10^{-6}	2.37×10^{-6}	2.23×10^{-6}	1.28×10^{-6}	2.79×10^{-6}	3.43×10^{-5}	2.84×10^{-5}
^{40}Ca	2.53×10^{-3}	5.29×10^{-3}	5.11×10^{-3}	3.25×10^{-3}	6.15×10^{-3}	2.56×10^{-2}	2.14×10^{-2}
^{44}Ca	1.22×10^{-4}	7.49×10^{-5}	1.43×10^{-5}	9.15×10^{-5}	2.11×10^{-5}	2.00×10^{-5}	2.97×10^{-4}
^{46}Ti	2.56×10^{-6}	6.26×10^{-6}	6.72×10^{-6}	6.81×10^{-6}	6.84×10^{-6}	3.56×10^{-5}	1.44×10^{-5}
^{47}Ti	5.13×10^{-6}	3.75×10^{-6}	3.11×10^{-7}	1.73×10^{-6}	9.11×10^{-7}	9.74×10^{-7}	6.26×10^{-7}
^{48}Ti	1.68×10^{-4}	1.58×10^{-4}	8.59×10^{-5}	1.85×10^{-4}	8.98×10^{-5}	1.58×10^{-4}	1.42×10^{-4}
^{49}Ti	3.45×10^{-6}	6.10×10^{-6}	7.54×10^{-6}	4.89×10^{-6}	6.01×10^{-6}	2.17×10^{-5}	6.97×10^{-6}
^{50}Ti	3.56×10^{-10}	1.21×10^{-9}	1.17×10^{-10}	1.12×10^{-10}	1.12×10^{-10}	2.00×10^{-10}	2.56×10^{-10}
^{50}Cr	2.30×10^{-5}	5.15×10^{-5}	7.49×10^{-5}	3.54×10^{-5}	5.01×10^{-5}	1.49×10^{-4}	1.01×10^{-4}
^{52}Cr	1.15×10^{-3}	1.36×10^{-3}	1.44×10^{-3}	8.64×10^{-4}	1.31×10^{-3}	2.77×10^{-3}	6.86×10^{-4}
^{53}Cr	9.34×10^{-5}	1.35×10^{-4}	1.50×10^{-4}	7.12×10^{-5}	1.39×10^{-4}	3.56×10^{-4}	1.00×10^{-4}
^{54}Cr	3.35×10^{-8}	4.09×10^{-8}	2.53×10^{-8}	6.26×10^{-9}	2.41×10^{-8}	2.81×10^{-8}	7.61×10^{-8}
^{55}Mn	3.65×10^{-4}	4.74×10^{-4}	5.48×10^{-4}	2.27×10^{-4}	5.02×10^{-4}	8.41×10^{-4}	3.64×10^{-4}
^{54}Fe	2.10×10^{-3}	4.49×10^{-3}	6.04×10^{-3}	2.52×10^{-3}	4.81×10^{-3}	9.17×10^{-3}	5.81×10^{-3}
^{56}Fe	1.50×10^{-1}	1.44×10^{-1}	7.57×10^{-2}	7.32×10^{-2}	5.24×10^{-2}	7.50×10^{-2}	7.50×10^{-2}
^{57}Fe	4.86×10^{-3}	4.90×10^{-3}	2.17×10^{-3}	3.07×10^{-3}	1.16×10^{-3}	2.29×10^{-3}	3.83×10^{-3}
^{59}Co	1.39×10^{-4}	1.22×10^{-4}	4.82×10^{-5}	1.31×10^{-4}	2.19×10^{-5}	2.51×10^{-5}	1.59×10^{-4}
^{58}Ni	5.82×10^{-3}	7.50×10^{-3}	3.08×10^{-3}	3.71×10^{-3}	1.33×10^{-3}	3.31×10^{-3}	9.25×10^{-3}
^{60}Ni	3.72×10^{-3}	3.36×10^{-3}	8.71×10^{-4}	2.18×10^{-3}	6.67×10^{-4}	3.88×10^{-4}	1.77×10^{-3}
^{62}Ni	1.05×10^{-3}	9.50×10^{-4}	2.52×10^{-4}	7.26×10^{-4}	1.70×10^{-4}	1.11×10^{-4}	1.28×10^{-3}

Source: Nomoto *et al.* (1997).

contribution to the general chemical enrichment of the to which society it belongs. Each star thus makes a donation according to its birth mass and metallicity.

The yields of the different stars are inscribed on a register painstakingly built up by a courageous generation of nuclear astrophysicists, including Stan Woosley, Ken Nomoto and Karl Friedrich Thielemann, to name the main contributors (see Table A4.1). This table of stellar yields and donations stands beside the abundance table of elements and isotopes as one of the treasures of human knowledge. However, the task is unfinished. According to its authors, it incorporates worrying uncertainties and defects. The table is

Table A4.2. *Masses of different elements ejected from a generation of stars*

Element	Ejected mass, $M_\odot$	
	Nomoto–Thielemann	Woosley–Weaver
C	7.93×10^{-2}	1.70×10^{-1}
N	1.56×10^{-3}	6.24×10^{-2}
O	1.80	1.17
F	1.16×10^{-9}	5.61×10^{-5}
Ne	2.31×10^{-1}	1.91×10^{-1}
Na	6.51×10^{-3}	4.81×10^{-3}
Mg	1.23×10^{-1}	7.08×10^{-2}
Al	1.48×10^{-2}	9.7×10^{-3}
Si	1.22×10^{-1}	1.30×10^{-1}
P	1.21×10^{-3}	1.59×10^{-3}
S	4.12×10^{-2}	5.94×10^{-2}
Cl	1.20×10^{-4}	6.21×10^{-4}
Ar	7.99×10^{-3}	1.16×10^{-2}
K	6.74×10^{-5}	3.28×10^{-4}
Ca	5.87×10^{-3}	6.86×10^{-3}
Sc	2.29×10^{-7}	3.88×10^{-6}
Ti	1.32×10^{-4}	2.10×10^{-4}
V	1.00×10^{-5}	2.65×10^{-5}
Cr	1.32×10^{-3}	1.70×10^{-3}
Mn	3.86×10^{-4}	5.88×10^{-4}
Fe	9.07×10^{-2}	1.14×10^{-1}
Co	7.27×10^{-5}	4.53×10^{-4}
Ni	5.97×10^{-3}	1.17×10^{-2}

ceaselessly updated and corrected as theoretical progress is made and further constraints imposed by observation. Nevertheless, in its current form, it can already be used to follow the main channels of galactic evolution.

Most helium, carbon, nitrogen and minor isotopes of carbon and oxygen such as carbon-13 and oxygen-17 and 18, together with heavy nuclei ($A > 100$) generated by the s process, originate in intermediate mass stars (2–$8\ M_\odot$). These are precisely the stars occurring on the asymptotic giant branch of the HR diagram.

Massive stars are the production and assembly line for most nuclear species. Intermediate elements, from carbon to calcium, are mainly produced by hydrostatic combustion, which is slow and non-explosive, one might say gentle. Iron on the other hand, together with its nuclear neighbours, arises in the final explosion (SNII), and also in type Ia supernovas. The amount of each newly synthesised nuclear species ejected from each star, usually referred to as the yield, can be calculated individually or collectively.

The results of the two groups Thielemann–Nomoto and Woosley–Weaver do converge globally speaking, but differ in the details (see Table A4.2). This is because they use different reaction probabilities and different algorithms for the mathematical treatment of convection within stars.

It is worth noting in passing that the ratios O/Fe, Mg/Fe, Si/Fe, Ca/Fe and Ti/Fe in the ejected matter are roughly three times greater than their solar counterparts. These excesses of α nuclei are observed in ancient stars of the galactic halo, suggesting that the explosion of massive stars (type II supernovas) may have produced them (see Chapter 8).

Values for the yields are affected by combined uncertainties in the basic physics of nuclear reactions and convection effects, during both pre-supernova evolution and the explosion itself. Nuclear contributions from stars depend primarily on the following:

- the nuclear reaction rates adopted and their temperature variations;
- the way convection is treated, along with the various mixing processes at work within the stars;
- the detailed treatment of the explosion and in particular the dividing line between the collapsing core and the ejected envelope, in the case of gravitational-collapse supernovas.

We have to deal with two distinct sets of problems concerning (a) nuclear reaction rates, and (b) the mathematical treatment of convection. However, their effects are combined in the result. As an example, combined nuclear and convective uncertainties affect the size of the carbon + oxygen core resulting from helium fusion as well as the ratio of carbon to oxygen within it. From there, they influence the ratio of the ashes of these elements and the mass of the iron core, which is a determining factor in the explosion.

The three problems, nuclear, convective and Gordian, are quite different in character and in gravity. Nuclear data are gradually being refined, thanks to a huge scientific effort, even though the crucially important reaction

$$^4\text{He} + {}^{12}\text{C} \longrightarrow {}^{16}\text{O} + \gamma$$

remains a thorn in the side of nucleosynthesis, given the great difficulty involved in measuring it.

The dividing line between core and envelope determines the amount of nickel-56 ejected. At the end of the day, this is what fixes the brightness of the supernova. We therefore have a handle on this parameter and for the time being may content ourselves by fitting it in such a way as to account for observations of supernova light curves.

But of the three problems, the most formidable is undoubtedly convection. An adequate treatment of the various mixing processes is the most important problem in the theory of stellar evolution. This is all the more upsetting in that the relevant theory appears to have reached a standstill. So, stellar stirrers, we hope to hear from you soon!

The yield table thus serves as a basis for modelling the chemical evolution of our Galaxy, or any other galaxy. Three distinct components must be specified: the yields of massive stars (8–100 $M_\odot$), which become type II supernovas, those of intermediate-mass stars (1–8 $M_\odot$), which blossom into planetary nebulas, and finally, those of overfed white dwarfs, which give birth to type Ia supernovas.

Appendix 5

Galactic evolution

Glossary

initial mass function mass distribution of stars at birth

Chemical evolution of galaxies

The key word in modern theory is 'evolution'. The impressive consistency of the astro-nuclear view of the heavens has established the idea of an evolution of nuclear species which has the same significance for astrophysics as the evolution of living species for biology. It is itself preceded by an evolution of particle or corpuscular species, which would have been very short, lasting less than 1 second. This process was of a quite crucial nature in determining the components available to build up atoms, that is, those stable particles, protons and neutrons, that serve as the building-blocks, and the forces that bind them together.

Once the elementary particles are produced, nuclear evolution precedes and determines all others, including geological and biological evolution, and its main agent is the stars. There are four main arguments to support the idea of a stellar genealogy for atomic matter. These can be described as the poverty of the ancients, the evolutionary trail, the great galactic cycle, and stellar alchemy. They are not independent of one another. Quite the contrary, they are very deeply related through the dialectic between big and small, astronomical and nuclear.

Note that I do not say 'infinitely small', for there are things smaller than atomic nuclei, namely elementary particles. There are also things larger than the astronomical scale of stars and galaxies that concerns us here.

The poverty of the ancients

Over the last half century, an astronomical fact of the greatest significance has been established with certainty: the chemical composition of the stars in our Galaxy and other galaxies undergoes variations and furthermore the age and heavy-element content, or metallicity, are inversely proportional. This strongly suggests that the Big Bang cannot be the origin of all the elements.

This conclusion eliminates Gamow's proposal according to which the temperature and pressure at the very beginning of the Universe would have been sufficient to produce elements beyond lithium. The hard reality of observations thus put paid to one of the most beautiful theories, attributing a common origin to all the elements.

The proportion of heavy elements varies by a factor of a thousand and more between stars of different ages and also depends on the level of development of the stellar society in question. As a general rule, in the kingdom of stars, the most ancient stars are also the least well provided for in metals. Hence, the galaxies grow richer as wave after wave of stellar workers come and go. The most recent generations of stars are richer than the ones that went before, and as a result of their labours. Each generation results in the confection of a whole range of nuclei and sees to their distribution through space via stellar winds, envelope ejection or explosion, thus making them available for further use.

The evolutionary trail

The trail followed by each star is inked onto the parchment of the HR diagram according to its mass and, to a lesser extent, its metallicity at birth. Each position on a particular evolutionary trail corresponds to a specific cycle of thermonuclear fusion. Every star follows a different path and at a different rate.

The configuration of the HR diagram is explained in terms of series of nuclear-fusion cycles. For example, the main sequence corresponds to hydrogen fusion, red giants to core helium burning, the asymptotic giant branch to shell hydrogen and helium burning.

The reddening of stars with age is a sign that they have exhausted their stock of hydrogen in the core. Gravitational contraction of the helium core is accompanied by shell hydrogen burning around the central region, whilst the temperature in the centre continues to rise. The star thus begins to ascend the giant branch. This was understood through the work of Chandrasekhar, Sandage and Schwarzschild between 1942 and 1952.

Ageing stars thus turn red, except for the most massive amongst them, which go purple or even ultraviolet. At the same time they move away from the main sequence. Their central temperature increases, as does their pressure, until further nuclear reactions are triggered which are capable of building up carbon from helium. This happens right through the long climb up the giant branch. Stars shine because they transmute the elements.

The construction of nuclear species in massive stars reaches its apotheosis in the explosion of supernovas.

The great galactic cycle

The great cycle of inspiration and expiration of matter, passing through condensation, nucleosynthesis, ejection and back to condensation, serves as a bellows to fire today's studies of galactic evolution. We owe this understanding to Fred Hoyle, present at every stage in the edification of our astro-nuclear doctrine.

This scheme merely illustrates the idea that the world is evolving everywhere, in every galaxy, and that the driving force behind cosmic evolution lies in the stars, born of gas clouds. This isolationist and private view (one might almost say maternal), contrasting with the global (or paternal) view of an expanding universe, singularises each star society in some sense, treating it as a unit well separated from others of its kind, evolving through its own initiative and at the rhythm of its reproduction.

The recycling of matter incinerated and transformed by stars and the gradual enrichment of matter into heavy elements as it passes from one stellar crucible to the next is the great scheme which forms the basis for the nuclear evolution of galaxies.

Stellar alchemy

A detailed demonstration of the fact that nucleosynthesis can occur inside stars across the whole range of atomic masses underpins everything that has been said so far. Stellar nucleosynthesis gained a great deal of credibility when it was shown that the Big Bang would never be able to circumvent the absence of a stable nucleus with mass 5.

The first major developments are attributed to Öpik as early as 1938 and independently to Salpeter in 1952 and then Hoyle. Salpeter and Hoyle showed that a triple α reaction would be capable of generating carbon inside stars. The leap across the chasm separating helium from carbon profoundly changed the field of nucleosynthesis, opening the way to nuclear complexity. Indeed, starting from carbon, the trail leads right up to uranium.

The demonstration that stars are capable of such nuclear fertility is based upon a combination of knowledge from what appear to be widely separated areas of physics. One of these concerns the internal structure of stars, telling us the temperature and pressure at different depths. The other concerns the probabilities at different energies of all the possible reactions between various nuclei, and between those nuclei and protons or neutrons. In the latter case, the acquisition of the relevant data was greatly accelerated by the Second World War. The beauty of nuclear astrophysics rests upon the success of this marriage and the complementarity of the two disciplines it brings together. The nuclear butterfly has returned to its stellar chrysalis.

The chemical diversity of the world as reflected in Mendeleyev's periodic table and the elemental abundance table finally has its explanation. Today we have reached the stage where we can study the chemical evolution of the galaxies with a view to writing the story of each element in its own astrophysical context. Let us not forget the name of Beatrice Tinsley, who initiated this vision, but died too soon to witness its achievement.

This confluence between two great streams of thought, the physics of the nucleus and the physics of stars, founds the notion of cosmic evolution which is as important for astronomy and cosmology as the evolution of the species for biology.

Is it perhaps because revolution has failed on Earth that humankind has set out to find evolution in the sky? Since that moment of revelation, we have been proclaiming that we are stardust, which is only partly true, because hydrogen is the ash of the Big Bang. Anyway, the cosmos is evolving. We are living in the golden age of evolutionary astrophysics.

A mental image of the evolution of the Galaxy

The logical foundation of the model is as follows: apart from the lightest elements, the history of the other elements in the Galaxy is dominated by nucleosynthesis in many generations of stars. Each one has a different history to all the others, at least, on the face of things. However, species with a common origin, for example, those produced abundantly by such and such a type of star, are likely to evolve in parallel. The picture we should imagine, or rather model, is thus relatively simple. It is based upon the following irreversible transformation rules, which define the direction of evolution:

$$gas \longrightarrow star,$$
$$simple\ nuclei \longrightarrow complex\ nuclei,$$
$$gas\ without\ stars \longrightarrow stars\ without\ gas.$$

The model assumes that evolution takes place in a closed system, with successive generations of stars being born into the interstellar medium. At each generation, a fraction of the gas is transformed into metals and returned to the interstellar medium. Gases imprisoned in stars of low mass and compact residues play no further role in galactic evolution. In this model, metallicity is bound to increase as time goes by. And so the arrow of galactic time is defined. Evolution will continue until no further gas is available to form new stars.

Apart from the lifetime of stars as a function of their initial mass, the galactic evolution kit contains the following three items:

- the production of heavy elements by stars of various masses, referred to as the nuclear yield;
- the mass distribution of stars or initial mass function;
- the star formation rate.

Amongst these three ingredients, only the first can yet be calculated from first principles, using the model of stellar evolution forged over the last few decades.

Yield and lifetime are individual properties, varying from star to star. The transition from individual to society involves knowledge of two demographic parameters, if we may put it that way. One concerns the distribution of stellar masses at birth within a single generation, the other the rate at which stars form, whatever their mass, at different periods in the life of the Galaxy.

The second parameter, the initial mass function, serves to weight the contributions of stars with different masses in proportion to their number within a single generation. The initial mass function has been established empirically and appears to remain fairly stable in time. The number of stars of mass M is inversely proportional to the cube of M, to a first approximation, provided we exclude the slightest of them ($M < M_\odot$). Looking at the mass distribution at birth, once established, we notice immediately how rare the massive stars are. For every star born at $10\ M_\odot$, there are a thousand births of solar-mass stars.

Assuming that the initial mass function is invariable, we may calculate the average production of the various star generations, born with the same metallicity, and estimate their contribution to the evolution of the galaxy (see Appendix 4). The abundances produced by a whole population are not as discontinuous and irregular as those shown in the table of individual yields (Table A4.1). This is because the latter are averaged over the mass distribution.

The third parameter in chemical evolution models of the galaxy, the star formation rate, serves to define the rate at which evolution proceeds. Unfortunately, there is a great deal of uncertainty in its estimation. There is no theory of star formation worthy of the name. In decline as time goes by, the star formation rate is often assumed to be proportional to the gaseous fraction of the Galaxy, which is itself in permanent decline, or to some power of it, but less than 2.

This parameter is nevertheless constrained by the relationship between the age and iron content of stars, the observed rate of supernova events (of the order of three per century), which is related to the current star formation rate, and the present gaseous fraction. After 10 billion years of evolution, the region of the Galaxy accessible to us, that is, the solar neighbourhood, still retains about 10% gas.

It is assumed that the region to which calculations apply is not influenced by any external matter supply and also that it suffers no loss of substance. In other words, it is assumed to operate in an autarkic manner, as in a closed box. Despite such grossly simplifying hypotheses, it is satisfying to watch the whole of galactic evolution unfurl before us in the form of such simple diagrams, provided of course that we are happy to abide by these strident directives. The model can be complicated at will by allowing the parameters to vary in time (or with metallicity, taken as the measure of time).

The whole art in the study of galactic evolution is to put forward a model that relates to available data, taking their volume and accuracy into account. Consequently, the evolutionary model must itself be conceived in an evolutionary way.

At the present level of our understanding, the simplest model seems good enough to explain the main trends deduced from systematic observations of stars belonging to different populations, provided that we exclude the period when the Galaxy itself was forming, which remains rather obscure. However, this picture will not be adequate in the future. More sophisticated numerical simulations are under investigation. These include hydrodynamic aspects of the chemical evolution of the Galaxy, which may be turbulent and violent, as well as its luminous or photometric counterpart. The results can then be related to spiral galaxies like our own.

However, let us return to the simple and robust model provided by the closed box idea and see how it works. Technically, physical quantities are calculated in a step-by-step manner. As an example, let us consider the quantity of gas present in a vast region of our Galaxy centred on the Sun. We shall also be concerned with its composition. At the first step in the waltz, the Galaxy is made entirely of gas and quite devoid of any complex nuclei. It has the composition attributed to it by the Big Bang. At the nth time step, of thickness dt, we examine the accounts to determine what has been gained and what has been lost in the interstellar medium over the period dt. The newly appearing nuclei were made inside stars that formed during previous time steps. The part removed from the gaseous medium is the part transformed into stars during the time step under consideration. We then move on to the next step and the process continues. One small step follows another and, starting from the most rudimentary matter, we finally arrive at our own magnificent Solar System and the present time.

On the profit side of the account, we carry over all nuclei ejected by stars at the end of their lives. This includes all those stars born in earlier epochs and entering the throes of death precisely at the time t in question. The exact amounts of nuclei depend of course on the mass of the dying star. Thus a star of mass M which dies at time t was born x years before, where x is the mass-dependent lifetime. For example, the nuclear donation, that is to say, the nuclear return on investment, from a star weighing in at 20 solar masses is made 10 million years after its birth, when it explodes. The return from a type Ia supernova occurs much later, at least 100 million years after the formation of a stellar couple with explosive vocation in which one of the members will eventually become a white dwarf. Even more extreme is the delivery date for stars with similar mass to the Sun. Those which formed at the beginning of the Galaxy are only just opening up today to spread their substance and give birth to those beautiful phenomena we know as planetary nebulas.

Concerning gas losses, we must subtract gas transformed into stars and the matter imprisoned in stellar corpses. The latter occur in three forms: white dwarfs, neutron stars and black holes. We must also include matter going into planets and aborted stars (brown dwarfs), forever frozen and permanently withdrawn from the (nuclear) chemical evolution of the Galaxy.

On the profit side of the account, we must record the gas rejected by each dying star, be it supernova or planetary nebula, and reinjected into the great galactic cycle.

The calculation is in principle quite straightforward, but the parameters are uncertain, as we mentioned earlier. For this reason, the theory of galactic evolution should not be viewed as a mature theory, but rather as a realistic scenario. Running the history of the Milky Way up to the present epoch, we can follow through the behaviour of gas, stars and metals, which are dialectically related. From this great material adventure, the following trends are singled out:

* growth and stabilisation of metallicity;
* decline in the star formation rate;
* reduction in the number of supernovas.

Considered as a natural object, the Galaxy is wearing out, for nothing is eternal. The golden age of nucleosynthesis is coming to a close. In our own region of the Milky Way, the stellar baby boom is a thing of the past. The great river of stellar births is drying up, because the amount of gas remaining is only a modest fraction of the total, maybe some 10%. The same is true for every other region of the Galaxy, which has exhausted its clouds, trading them for stars. An exception may perhaps be found in a gaseous ring located at a radius of about 13 000 light-years from the galactic centre, where many star births are still celebrated.

The low rate of supernovas is also witness to a considerable slowing down in star formation activity, for want of fuel. Those few supernovas still happening will barely modify the current composition of the interstellar medium. The chemical evolution of the galactic disk is therefore coming to an end. It will result in a general stabilisation of abundances at levels similar to those in the Sun. The Sun is mortal! The Galaxy is mortal! This is a poignant moment in the story of our Galaxy, when it burns its last fires. It is fading, but we shall fade long before it.

Appendix 6

Key dates

1572	Brahe	Supernova
1610	Galileo	Refracting telescope
1731	Messier	Catalogue of nebulas
1916	Einstein	Relativistic universe, static solution
1917	de Sitter	Relativistic universe without matter
1922	Friedmann	Relativistic universe, dynamic solution
1924	Hubble	Discovery of Cepheids in Andromeda
1933	Hubble, Humason	Recession of the galaxies ($v = Hd$)
1933	Lemaître	Notion of the beginning of the Universe
1939	Bethe	Hydrogen fusion in stars
1948	Gamow	Prediction of cosmological background radiation
1948	Gamow, Bethe, Alpher	Primordial nucleosynthesis
1957	Burbidge, Burbidge, Fowler, Hoyle, Cameron	Stellar nucleosynthesis
1965	Penzias, Wilson	Discovery of cosmological background radiation
1999	Perlmutter, Riess, Smith	Acceleration of the expansion of the Universe

Appendix 7

Constants and units

Universal constants

Speed of light in vacuum	$c = 2.99792458 \times 10^{10} \text{ cm s}^{-1}$
Gravitational constant	$G = 6.67259 \times 10^{-8} \text{ cm}^3\text{g}^{-1}\text{s}^{-2}$
Planck constant	$h = 6.62620 \times 10^{-27} \text{ erg s}$

Electron, proton, neutron

Electron charge	$e = 4.803242 \times 10^{-10} \text{ esu}$
	$e = 1.60217733 \times 10^{-19} \text{ coulomb}$
Electron mass	$m_e = 9.1093897 \times 10^{-28} \text{ g}$
Electron mass energy	$m_e c^2 = 0.5109906 \text{ MeV}$
	$m_e c^2 = 8.1871 \times 10^{-7} \text{ erg}$
Classical electron radius	$r_e = 2.81794092 \times 10^{-13} \text{ cm}$
Compton wavelength of the electron	$\lambda_e = 2.42631058 \times 10^{-10} \text{ cm}$
Proton mass	$m_p = 1.6726231 \times 10^{-24} \text{ g}$
Neutron mass	$m_n = 1.6749286 \times 10^{-24} \text{ g}$

Physicochemical constants

Avogadro's constant	$N_A = 6.022137 \times 10^{23} \text{ mol}^{-1}$
Boltzmann's constant	$k_B = 1.380658 \times 10^{-16} \text{ erg K}^{-1}$

Astronomical constants

Light-year	$1 \text{ ly} = 9.46530 \times 10^{17} \text{ cm}$
Parsec	$1 \text{ pc} = 3.085678 \times 10^{18} \text{ cm}$
	$1 \text{ pc} = 3.261633 \text{ ly}$
Solar mass	$M_\odot = 1.9891 \times 10^{33} \text{ g}$
Solar radius	$R_\odot = 6.9599 \times 10^{10} \text{ cm}$
Solar luminosity	$L_\odot = 3.8268 \times 10^{33} \text{ erg s}^{-1}$
Effective temperature of the Sun	$T_{\text{eff}} = 5780 \text{ K}$

Appendix 8

Websites

A great deal of information is available on the World Wide Web concerning the large observatories on Earth and in space.

ALMA	www.eso.org/projects/alma
ESA	www.esa.int
ESO	www.eso.org
FUSE	fuse.pha.jhu.edu/
INTEGRAL	sci.esa.int/home/integral/index.cfm
Kueyen–UVES	www.eso.nce.ut2sv
MAP (Microwave Anisotropy Probe)	map.gsfc.nasa.gov
NASA	www.nasa.gov
NGST	ecf.hq.eso.org/ngst/ngst.html
SNAP (Supernova/Acceleration Probe)	snap.lbl.gov
VLT	eso.org
XMM	sci.esa.int/xmm/

Bibliography

Allègre C. (1996) *De la pierre à l'étoile*, 2nd edn. (Fayard, Paris).

Anders E. & Grevesse N. (1989) 'Abundances of the elements: meteoritic and solar.' *Geochim. Cosmochim. Acta* **53**, 197.

Arnett D. (1996) *Supernovae and Nucleosynthesis* (Princeton University Press, Princeton).

Arnould M. & Takahashi K. (1999) 'Nuclear astrophysics.' *Rep. Prog. Phys.* **62**, 393.

Audouze J. (1998) 'L'Univers'. *Que sais-je?* (Presses Universitaires de France, Paris).

Audouze J. & Cazenave M. (2000) *L'Homme dans ses univers* (Albin Michel, Paris).

Audouze J. & Israël G. (eds.) (1983) *The Cambridge Atlas of Astronomy* (Cambridge University Press, Cambridge).

Audouze J. & Vauclair S. (1998) 'L'Astrophysique nucléaire.' *Que sais-je?* (Presses Universitaires de France, Paris).

Audouze J., Cassé M. & Carrière J.C. (1998) *Conversations sur l'invisible* (Plon, Paris).

Audouze J., Musset P. & Paty M. (1990) 'Les Particules et l'univers.' *Nouvelle Encyclopédie Diderot* (Presses Universitaires de France, Paris).

Bahcall J.N. (1989) *Neutrino Astrophysics* (Cambridge University Press, Cambridge).

Bahcall J.N. (2002) *Solar Models: An Historical Overview*. CERN Document Server astro-ph/0209080.

Barrow J. (1991) *Theories of Everything* (Oxford University Press, Oxford).

Brahic A. (1999) *Enfants du soleil* (Odile Jacob, Paris).

Brown G., Kamionkowski M. & Turner M. (2000) 'David Schramm's Universe.' *Phys. Rep.* **6**, 333.

Burbidge M., Burbidge G., Fowler W. & Hoyle F. (1957) *Rev. Modern Phys.* **29**, 647.

Burles S.M. & Tytler D. (1997) 'Closing in on the primordial abundance of deuterium.' CERN Document Server astro-ph/9712265.

Cameron A.G.W. (1957) *Atomic Energy of Canada*, Ldt, CRL-41.

Cassé M. (1987) *Nostalgie de la lumière* (Belfond, Paris).

Cassé M. (1993) *Du vide et de la création* (Odile Jacob, Paris).

Cassé M. (1998) *Nucléosynthèse et abondance dans l'univers* (Cépaduès, Paris).

Cassé M. (1999) *Théories du ciel* (Payot, Paris).

Cazenave M. (1999) *Dictionnaire de l'ignorance* (Albin Michel, Paris).

Clayton D.D. (1983) *Principles of Stellar Evolution and Nucleosynthesis* (University of Chicago Press, Chicago).

Cribier M., Spiro M. & Vignaud D. (1995) *La Lumière des neutrinos* (Le Seuil, Paris).

Crozon M. (1999) *L'Univers des particules* (Le Seuil, Paris).

Daniel J.Y. (1999) *Sciences de la terre et de l'univers* (Vuibert, Paris).

Davies P.C.W. (1989) *The New Physics* (Cambridge University Press, Cambridge).

Doom C. (1986) *La Vie des etoiles* (Le Rocher, Paris).

Friedmann A. & Lemaître G. (1997) *Essais de cosmologie*, preceded by *L'Invention du Big Bang* by J.P. Luminet (Le Seuil, Paris).

Kaler J.B. (1997) *Stars and their Spectra* (Cambridge University Press, Cambridge).

Klein E. & Lachièze-Rey M. (1996) *La Quête de l'unité* (Albin Michel, Paris).

Kolb E. & Turner M.S. (1990) *The Early Universe* (Addison-Wesley, New York).

Krauss L. (1989) *The Fifth Essence: The Search for Dark Matter in the Universe* (Basic Books, New York).

Lachièze-Rey M. (1987) *Connaissance du cosmos* (Albin Michel, Paris).

Lachièze-Rey M. & Gunzik E. (1995) *Le Fond diffus cosmologique* (Masson, Paris).

Lattimer J. & Burrows A. (1988) 'Neutrinos from Supernova 1987A.' *Sky and Telescope*, October 1988.

Lehoucq R. & Cassé M. (1990) In *Supernovae, Les Houches 1990* (North Holland, Amsterdam).

Léna P., Lebrun F. & Mignard M. (1998) *Observational Astrophysics* (Springer-Verlag, Berlin).

Longair M.S. (1997) *High Energy Astrophysics*, vol. 1, *Particles and their Detection*, vol. 2, *Stars, the Galaxy and the Interstellar Medium* (Cambridge University Press, Cambridge).

Luminet J.P. (1992) *Black Holes* (Cambridge University Press, Cambridge).

Luminet J.P. (1996) *Les Poètes et l'univers* (Le Cherche Midi Editeur, Paris).

Mochkovitch R. (1994) 'An introduction to the physics of type II supernova explosions', in *Matter under Extreme Conditions* (Springer-Verlag, Berlin).

Narlikar J. (1988) *The Primordial Universe* (Oxford University Press, Oxford).

Nomoto K. *et al.* (1997) 'Nucleosynthesis in type Ia supernovae.' CERN Document Server astro-ph/9706025.

Nomoto K. *et al.* (2001) 'Gamma-ray signatures of supernovae and hypernovae.' CERN Document Server astro-ph/0110079.

Nottale L. (1994) *L'Univers et la lumière* (Flammarion, Paris).

Ostlie D.A. & Carrol B.W. (1996) *Modern Stellar Astrophysics* (Addison-Wesley, New York).

Pagel B.E.J. (1997) *Nucleosynthesis and Chemical Evolution of Galaxies* (Cambridge University Press, Cambridge).

Pasachoff J.M. (1977) *Contemporary Astronomy* (W.B. Saunders, Philadelphia).

Paul J. (1998) *L'Homme qui courait derrière son étoile* (Odile Jacob, Paris).

Peebles P.J. (1993) *Principles of Physical Cosmology* (Princeton University Press, Princeton).

Perlmutter S. *et al.* (1999) *Phys. Rev. Lett.* **83**, 670.

Phillips A.C. (1994) *The Physics of Stars* (John Wiley, New York).

Prantzos N. & Montmerle T. (1998) 'Naissance, vie et mort des étoiles.' *Que sais-je?* (Presses Universitaires de France, Paris).

Prantzos N., Vangioni-Flam E. & Cassé M. (1993) *Origin and Evolution of the Elements* (Cambridge University Press, Cambridge).

Ramaty R., Vangioni-Flam E., Cassé M. & Olive K. (1997) 'Lithium, beryllium, boron, cosmic rays and related X and gamma rays.' *Astronomical Society of the Pacific Conference Series* **171**.

Reeves H. (1991) *Hour of our Delight: Cosmic Evolution, Order and Complexity* (W.H. Freeman, New York).

Reeves H. (1994) *Atoms of Silence* (General Publishing, Toronto).

Reeves H. (1996) *Last News from the Cosmos* (General Publishing, Toronto).

Ricard M. & Thuan T.X. (2000) *L'Infini dans la paume de la main* (Fayard, Paris).

Riess A.G. *et al.* (1998) 'Light curves for 22 type Ia supernovae.' CERN Document Server astro-ph/9810291.

Riordan E.M. & Schramm D.N. (1992) *The Shadows of Creation: Dark Matter and the Structure of the Universe* (W.H. Freeman, New York).

Rolfs C.E. & Rodney W.S. (1998) *Cauldrons in the Cosmos: Nuclear Astrophysics* (University of Chicago Press, Chicago).

Schatzman E. (1986) *Les Enfants d'Uranie* (Le Seuil, Paris).

Schatzman E. & Praderie F. (1990) *Les Étoiles* (Interédition, Paris).

Schönfelder V. (ed.) (2001) *The Universe in Gamma Rays* (Springer-Verlag, Berlin).

Sexl R. & Sexl H. (1979) *White Dwarfs – Black Holes: An Introduction to Relativistic Astrophysics* (Academic Press, New York).

Shu F.H. (1982) *The Physical Universe* (University Science Books, New York).

Silk J. (1994) *A Short History of the Universe* (Scientific American Library, New York).

Silk J. (1997) *The Big Bang*, 2nd edn. (W.H. Freeman, New York).

Slezak E. & Thévenin F. (1998) *Nucléosynthèse et abondance dans l'univers* (Cépaduès, Paris).

Sneden C. (2001) 'Neutron-capture element abundances and cosmochronometry.' CERN Document Server astro-ph/0106366.

Thielemann F.K. *et al.* (2001) 'Element synthesis in stars.' CERN Document server astro-ph/0101476.

Thuan T.X. (1988) *La Mélodie secrète* (Fayard, Paris).

Thuan T.X. (1998) *Le Destin de l'univers* (Gallimard Découverte, Paris).

Tubiana R. & Dautray R. (1996) 'La Radioactivité et ses applications.' *Que sais-je?* (Presses Universitaires de France, Paris).

Tytler D. (1997) 'Cosmology with the Very Large Telescope Interferometer using a space-based astrometric reference frame.' CERN Document Server astro-ph/9701197.

Valentin L. (1982) *Physique subatomique: noyaux et particules*, vols. 1 and 2 (Hermann, Paris).

Vangioni-Flam E. & Cassé M. (1998) In *La Recherche*, Special Issue no. 1, April 1998.

Vangioni-Flam E. & Cassé M. (1999) *Petite Étoile* (Odile Jacob, Paris).

Vauclair S. (1996) *La Symphonie des étoiles* (Albin Michel, Paris).

Verdet J.P. (1993) *Astronomie et astrophysique* (Larousse, Paris).

Vidal-Madjar A. (2000) *Il pleut des planètes* (Hachette, Paris).

Weinberg S. (1993) *The First Three Minutes* (Basic Books, New York).

Woosley S. & Weaver T. (1989) 'The great supernova of 1987.' *Scientific American*, August 1989.

Index

237